ECOLOGY
of
INSECT VECTOR
POPULATIONS

ECOLOGY
of
INSECT VECTOR
POPULATIONS

R. C. Muirhead-Thomson

*London School of Hygiene
and Tropical Medicine, London, England*

Academic Press · 1968
London and New York

ACADEMIC PRESS INC. (LONDON) LTD
BERKELEY SQUARE HOUSE
BERKELEY SQUARE
LONDON, W.1

U.S. Edition published by
ACADEMIC PRESS INC.
111 FIFTH AVENUE
NEW YORK, NEW YORK 10003

Printed in Great Britain by
W & J Mackay & Co Ltd, Chatham, Kent

PREFACE

The present book is an attempt to review and appraise the basis of existing knowledge about the ecology of insect vectors of disease, with particular reference to the problems of sampling adult vector populations. The great bulk of information about the abundance, distribution and behaviour of the many different types of insect which transmit human and animal disease is based essentially on data provided by a variety of capture or sampling methods, and on the interpretation put on such data.

The review examines the manner in which the inter-related problems of insect ecology, behaviour and population sampling have been explored over a wide range of investigations on disease vectors. Particular attention has been given to the blood-sucking dipterous insects such as tse-tse flies, mosquitoes, black-flies, midges etc. but an appraisal has also been made of relevant studies on house-fly and blow-fly populations. The special problems encountered in studying ecology of the wingless vectors of disease have been exemplified by the extensive work carried out on fleas, not only in their role as vectors of plague, but also in relation to myxomatosis.

It is hoped that this comprehensive review, and its critical analysis of common problems and objectives, will be of interest not only to the medical entomologists and epidemiologists most directly involved, but also to a wider range of ecologists concerned with general problems of animal populations and the quantitative approach to animal ecology.

ACKNOWLEDGEMENTS

I am indebted to Dr. T. A. M. Nash for some very helpful comments on a very early draft of the chapter on tse-tse flies, and I am particularly grateful to Mr. W. H. Potts for corrections and suggestions made on the final draft of the same chapter. I am also grateful to Mr. J. Hadjinicolau for kindly allowing me to use his unpublished data on which Fig. 5 is based. To Mr. C. J. Webb and his staff of the Visual Aids Department of the London School of Hygiene and Tropical Medicine I am greatly indebted for the preparation of all the charts and histograms presented in Figs. 1–13.

May 1968   R. C. MUIRHEAD-THOMSON

CONTENTS

CHAPTER 1

INTRODUCTION

The intensive studies carried out in many countries on all the major vectors of insect-borne human and animal disease since the beginning of this century have produced a most impressive body of information about their habits, distribution, and role in disease transmission. Certain groups of these insects such as the mosquito vectors of malaria, the mosquitoes involved in sylvan or "jungle" yellow fever, and the tse-tse flies responsible for the transmission of human and animal trypanosomiasis have been the object of particularly intensive study by specialists in many disciplines (entomology, parasitology, ecology, and public health).

During the last ten years or so there has been a very great intensification of the attack against insect-borne disease, on an international scale as well as on a national one. Alongside this there has been a progressively increasing demand for more accurate information about the insect vectors' part in the complex ecology and epidemiology of the various diseases concerned. There has also been a demand for new and more precise quantitative data about vector densities, distribution and behaviour, not only with a view to placing vector control programmes on a much sounder scientific basis, but also to enable the effects of various types of control programme to be accurately assessed and evaluated.

In addition, the rapid changes in many developing countries have tended to intensify or create new public health problems in areas where existing knowledge of vectors involved may be quite inadequate. The introduction and development of entirely new potential methods of insect vector control involving sterile male techniques and genetic manipulation have also produced a sudden need for information about certain specialized aspects of vector ecology which have received very little attention in the past.

Although the accumulated knowledge of vector ecology is undoubtedly impressive by most standards, nevertheless these events in recent years have shown that there is certainly no reason for complacency. Information on many aspects of ecology still remains very patchy and incomplete, and even apparently well-established beliefs have on occasion proved to be inaccurate or misleading on critical re-examination.

The object of the present study is to review current knowledge about the ecology of insects which transmit disease, and in particular to make a critical assessment of the various methods developed by different groups of entomologists in their investigations on basic problems of distribution, abundance, ecology and behaviour of vectors. In the light of that review it may then be possible to see in what way the more progressive or positive features can be further developed to meet the continuing need for more accurate quantitative information on vector ecology. This exercise would also be well in keeping with current trends in animal ecology as a whole where there is increasing emphasis for more accurate and reliable data on population dynamics.

CHAPTER 2

GENERAL PROBLEMS OF VECTOR ECOLOGY

If we view the ecology of vector-borne disease as a whole, it appears enormously complex, not only with regard to the range of human diseases and relevant vector-borne zoonoses, but also with the range and variety of the insect vectors concerned, to say nothing of the range of other arthropods involved. In view of this complexity it is not surprising that work on vector-borne disease has for long been compartmentalized, some workers being entirely concerned with malaria, others with filariasis, with trypanosomiasis, with yellow fever and so on. Within each of these disciplines there have been further inevitable subdivisions, mainly on a geographical basis, with the result that some workers have devoted their working life to a study of, say, tse-tse in one single limited part of Africa, or to the malaria vectors peculiar to a single country or even limited part of that country. Naturally there are exceptions to this: some medical research workers in South-East Asia have moved with great success from study of malaria to that of filariasis, and to mosquito-borne viral disease, while in Brazil one group of workers has managed to investigate successfully and successively the vectors of malaria, filariasis, Chagas disease, leishmaniasis and, finally, the snail intermediate hosts of bilharziasis.

Nevertheless, the general tendency has been towards increasing specialization, with each worker finding that the business of doing his own specialized work and keeping up with his own specialized literature is a full-time occupation. Occasional attempts to take a look at the world outside one's own particular groove tend to be discouraging. For example, the newcomer who attempts to familiarize himself with the problem of tse-tse fly and African trypanosomiasis finds himself confronted with such a voluminous and complex literature and with such a bewildering ecological kaleidoscope of tse-tse, game, cattle, human habits, vegetation types, climatic differences and so on, that it is only too easy to be discouraged. In the same way malaria eradication on a world-wide basis now deals with such a range of conditions and problems, and has to adapt itself to such rapidly changing events, that it is becoming a highly specialized subject comprehensible only to those closely involved.

However, a moment's reflection will show that all these specialized disciplines have so many fundamental points in common, that there is no longer any real justification for each to pursue its independent course, uninfluenced and unaware of the other. Workers in each of these many fields of insect-borne diseases are all concerned with such basic considerations as identity of adult vectors and non-vectors, distribution of vectors, seasonal incidence of vectors, biting habits in relation to man, natural infection rates, capacity to acquire infection experimentally, methods of catching and trapping vectors, and the assessment of the efficacy of various control measures, including the use of insecticides. Although the workers on these entomological problems are entomologists, with the same basic entomological training, the studies in each specialized field tend to develop on different lines. The inevitable result of this is that some aspects of vector biology are more thoroughly investigated in one particular field, while in another field special progress may be made in another direction. The logical deduction from this is that each specialized field has disclosed some aspect of general vector biology which may be of considerable value in other fields. Conversely, each specific field is not so self-contained that it can afford to ignore relevant advances in allied fields.

Of all the various points in common to the study of insect vectors, perhaps the most important is the question of sampling populations of adults. The entomologists in all the fields mentioned above are concerned with the various ways in which they can find or catch their particular vectors, and in the way in which such catches can be adapted or standardized in order to provide a reasonably close estimate of change in the density of the vector population as a whole, according to season, according to habitat or geographical range, and according to various control measures applied. In addition the entomologist is concerned with the composition of his catch according to the stage of hunger cycle or ovarian development, proportion of young and old insects, proportion infected and infective, and so on. He is also concerned with the extent to which this arbitrary catch or sample is really representative of the adult population at large, or of a particular population in which he is interested. He may find that one particular catching method yields the highest catch, but is only representative of one section of the population. The interpretation of so many vector activities is closely bound up with the validity of the sampling methods used. Sampling confined mainly to a single convenient method may give rise to very erroneous ideas about blood feeding, choice of host, longevity, infection rate and so on.

In the past many of the methods which have been adopted for the routine capture or sampling of insect vector populations have been determined by the fact that in any representative area the distribution of these insects is liable to be extremely patchy, with high concentrations or aggregations in certain restricted environments such as a thicket, a human habitation, or in the vicinity of an animal host. In many cases practical dictates of time and labour have restricted sampling to one particular type of aggregation site — not always clearly defined — which happens to be the most convenient or easily accessible one in the circumstances.

The nature of the capture or sampling method used has also played a major part in studies on the behaviour of adult vectors under different conditions or at different periods. When we talk about studies on behaviour it is often assumed that this implies the continued observation of one insect or a group of insects rather after the manner of the old classical studies of Fabre and others. But as far as insect vectors of disease are concerned, ideas of behaviour have more often been based on indirect inference than on continued direct observation. In the case of nocturnal vectors such as most species of mosquito, or of elusive daytime vectors such as tse-tse flies, continued observation on single insects is extremely difficult under natural conditions in the field. In such cases ideas of behaviour have been inferred from a series of samples which may differ in total numbers or in composition, according to age, hunger condition, ovarian development, etc. According to these differences it is deduced that vectors of a certain physiological condition are only active at certain periods of the night, are or are not attracted to animal bait, and do or do not rest in certain places and so on. In the case of vectors like mosquitoes which may enter human habitations to feed or rest or both, such nocturnal movements cannot readily be observed directly, but can only be inferred from their movements in and out of the house as judged by the use of a variety of catching, sampling, or trapping methods. Clearly, if the design of these sampling methods is uncritical or biassed, it is only too easy to arrive at very erroneous ideas about vector activity and vector reactions.

The need for a reappraisal of these sampling methods is particularly urgent in the matter of interpreting the effects of various control measures, particularly control by insecticides. Sampling methods applicable under normal untouched conditions may be found to be quite untenable or unrealistic after the wide-scale application of insecticide. If new sampling methods have to be devised at this stage, the further question arises as to how far these new data are directly comparable to those obtained by older methods.

In the last few years many entomologists, particularly those working with tse-tse and with anopheline mosquitoes have become increasingly aware of all these difficulties and pitfalls, and increasingly critical of long-established basic sampling methods. For this reason prominence will be given in this survey to those two independent fields which have been pursuing parallel courses for so long with the minimum of exchange or interchange.

In a review of this kind in which a general appraisal must necessarily include developments outside one's own speciality, there are clearly certain pitfalls and limitations. The reviewer has to rely on such published information as is available, which usually describes work which was completed a year previously or usually more. In this necessarily selective study of the literature there is a real danger, too, of overlooking significant contributions which may have been made in the form of less accessible reports. It is hoped that due allowance will be made for these deficiencies in a general appraisal of this kind. For this reason, therefore, the present report should be regarded not as an exhaustive review or an olympian judgement on trends in different fields, but rather as a first attempt to draw the attention of different workers to common problems in vector ecology, and to introduce workers in one field to significant developments and significant literature in other fields.

The bulk of this review will be concerned with the important winged dipterous vectors of disease, namely mosquitoes, tse-tse flies, black-flies, house flies and others. The very different problems entailed in the sampling of wingless vectors, which spend a great deal of their lives in close association or attachment to their hosts, is exemplified by the flea which, in its role as vector of human and rodent plague — as well as murine typhus — has been the subject of a considerable amount of research since the beginning of this century. The general problem of flea sampling and flea behaviour will also be illuminated by the more recent studies on the rabbit flea as a vector of myxomatosis.

No attempt has been made in this review to deal with the problems of sampling immature stages of insect vectors. There is a great deal of scattered information about sampling mosquito larvae, eggs and pupae in various types of water body, about black-fly larvae in streams and rivers, about tse-tse pupae in soil, and about larvae of midges and sand-flies which live in mud or damp soil. An analysis of this particular aspect of sampling could only be done adequately on the basis of comparison with experience in allied fields, such as fresh-water biology, snail ecology, fishery research, and studies on soil fauna. With one or two notable exceptions there have not been a sufficient number of detailed studies on the vector side to justify a separate review, but changing demands may yet create the need for much more critical studies on larval populations and other immature stages of vector insects (Knight, 1964).

Although the subject of this present review may appear to be one of rather limited scope in the general context of entomology or animal ecology, it is hoped that the wider implications of this analysis and appraisal will not be overlooked. Fundamentally similar basic problems of population sampling in ecology are being encountered in a wide range of biological studies (Andrewartha, 1961; Andrewartha and Birch, 1960; Morris, 1960; Williams, 1964; Southwood 1966; Clark *et al.* 1967). Medical entomologists should be able to benefit to a much greater extent not only from each other, but also from the experience gained by investigators in agricultural entomology, in fishery research, and many other branches of animal ecology. It is hoped that the benefit will be reciprocal and that ecologists in general may learn something of advantage from this analysis of basic ecological methods.

CHAPTER 3

TSE-TSE FLIES

The study of tse-tse fly ecology in Africa is particularly instructive in that many of the classical investigations — carried out continuously over several years — have provided unusual insight into the complex relationship between vector behaviour and sampling vector populations in general. The approach to tse-tse ecology also provides an interesting contrast to parallel studies on the ecology of mosquitoes, black-flies, sand-flies, and midges, in that both male and female tse-tse are bloodsuckers and vectors of disease, and that consequently sampling and capture techniques have had to take both sexes into account. Many of these investigations in Africa have been concerned with tse-tse as vectors of animal trypanosomiasis, particularly of domestic stock, while others have been primarily devoted to tse-tse as vectors of human disease. This distinction does not affect the present review unduly as many of the tse-tse studied are capable of transmitting both human and animal trypanosomiasis, and essentially the same entomological methods have been used, or at least tried out, over a wide range of species. Several reviews which have appeared in the last few years provide an excellent coverage to the many varied aspects of tse-tse and disease, for example, the general problem of tse-tse and trypanosomiasis (Nash, 1960; Ashcroft, 1959), the ecology of the main vectors (Langridge *et al.* 1963), the distribution and abundance of tse-tse (Glasgow, 1963) and more recent fundamental work (Glasgow, 1967). The whole subject of tse-tse ecology and sampling methods was reviewed by Buxton (1955) in his masterly treatise which will long remain a classic.

TSE-TSE SAMPLING AND ECOLOGY

The subject of tse-tse sampling and ecology is a peculiarly difficult one to discuss impartially. Experiences of different workers tend to vary considerably depending on the geographical region in which they work, and on the particular species or groups of species which they are investigating. In this chapter there will naturally be emphasis on certain features or trends which appear to be of particular interest in the context of vector sampling as a whole. This emphasis should not be taken as indicating any partisanship, or as any claim to superior judgement about the merits of different methods in the definitive field of tse-tse biology.

There is one particular direction in which tse-tse workers have made an outstanding contribution towards a better understanding of vector populations in general, namely in the design of field experiments involving the capture, marking, release and recapture of adult tse-tse, and in the analysis of the data obtained. As the sampling and ecological problems involved in this method of investigation have much in common with the more recent work carried out on blow-fly populations, they will be discussed concurrently in a later chapter.

Of the many different sampling methods which have been used there are four which merit special mention.

1. The fly round
2. The use of stationary bait
3. Trapping
4. Capture of resting tse-tse

1. *The fly round*

In the fly round in its classic form as designed in East Africa mainly for those species of tse-tse which roam far and wide in savannah country, a path is laid out crossing a range of vegetation types. A group of fly boys who act as combined bait and catchers move along this path, stopping from time to time to collect the tse-tse which follow them or are attracted to them. Male tse-tse usually make up the bulk of this catch — up to 90%. In recording the data for the catch, the teneral males — that is, those which are recently emerged and are comparatively soft bodied — are rejected, and the catch is recorded as the "number of non-teneral males per 10 000 yd". This figure is often referred to as the "apparent density".

This method is designed particularly for those species of tse-tse which are attracted by human bait — e.g. *Glossina morsitans*. In certain cases it has been found that the use of a bait cow — rather than human bait — in the fly round is necessary to attract certain more elusive species such as *G. pallidipes*. For many species of tse-tse and for many parts of Africa this method has been found to be the most convenient, even though entomologists have long been conscious of its limitations and deficiencies. However, it has come in for considerable criticism and reappraisal in the last few years, and it would therefore be instructive to inquire more closely into the rationale of this sampling technique.

The first point is that, although both sexes of tse-tse are bloodsuckers, and disease vectors, the fly-round catch yields mainly males of the savannah species. From the nature of this method one would assume that the basic principle is the attraction of a hungry fly to attractive bait. However, the situation appears to be rather more complicated than this. Many of the males caught are not hungry, and it has therefore been assumed that they follow the bait or moving object as part of the general search for female tse-tse.

The second point is that, even in areas where all other evidence points to the total tse-tse population remaining stable for weeks on end, there are liable to be considerable day-to-day fluctuations in the fly-round catch. It is evident that this sampling method is not necessarily a measure of the true population, because conditions which reduce the activity of the fly, such as climatic changes, will reduce the chances of its moving within the sphere of influence of the fly boys. This activity factor, along with others, is referred to as the "availability" of the fly, and it is evident that the fly-round catch, or "apparent density", is in some way determined by the true population and by the availability at that particular time (Jackson, 1949, 1953, 1954). Among the other factors which go to make up "availability" is the diurnal cycle of biting activity whereby some tse-tse show greater feeding activity at certain times of the day. One can visualize that at very low tse-tse densities there would need to be some synchronization of fly round with peak activity.

Despite these freely admitted imperfections in the fly round method of sampling, as applied to the woodland savannah species of East Africa, it is still considered sufficiently approximate to indicate major alterations in the tse-tse population such as would be expected as a result of heavy insecticide pressure, for example in judging the effect of aircraft application of insecticide to tse-tse bush (Foster *et al.* 1961, Burnett *et al.* 1961, 1965). In the fly round used in the latter instances, an additional record is kept of females and teneral males.

One of the possible sources of error in the normal fly round is the human one involved in leaving it to fly catchers to decide exactly where to halt, and how long to spend at each place. An attempt to overcome this variable has been made (Ford *et al.* 1959, Glasgow, 1961) by introducing the idea of the "transect fly round". Instead of being divided into a few long sections according to the investigators' assessment of the vegetation, the new round follows arbitrary straight lines, and is divided into numerous short sections of equal length. The catching party halts at the posts defining the sections, and catching takes place only at these halts. This method appears to have advantages in being more standardized than the older method, and less exposed to human variables. Whether it can deal with the more basic variables of the fly round described above remains to be seen.

The Belgian workers have also been interested in further developing the basic fly round with a view to obtaining more representative samples of *Glossina morsitans* under a wide range of conditions in the Congo. After drawing attention to some of the recognized limitations of the conventional fly round, they have developed a sampling plan based on capturing flies along the four sides and two diagonals of a number of squares 100 m x 100 m, each square forming part of an assembly of squares within a rectangular area whose exact dimensions are determined by the variety and distribution of habitats. (Lambrecht, 1958; van den Berghe and Lambrecht, 1962) Flies captured in these rounds are marked and released in order to eliminate the chance of recording the same fly more than once. As the fly capturers carry out the collecting to a depth of 10 m on each side of the straight path constituting the sides and diagonals of each square unit, a high proportion of the total area of each square is actually sampled. It would appear *a priori* that this modification in sampling technique would provide a closer approach to real fly density; but although a trial was designed to compare the relative merits of the two different fly rounds under similar conditions, the opportunity for making a valid comparison between square fly round and linear fly round did not arise.

In the light of all the experience gained in tse-tse studies in Africa, these workers later proposed a new type of fly round designed to meet a wide range of requirements (van den Berghe and Lambrecht, 1963). This new type of transect fly round involved at least 3 rounds of 1 000 m, each round having a separate fly-boy team, and being subdivided into 50 m sections. One of these rounds would be close to human settlements and habitations; one would be laid out in typical fly habitat and game habitat, and one round would be designed to include possible fly/game concentration points during the dry season. Each round would be sampled during 2-hour periods four times a day, once a week, and thus provide additional data on activity at different times of the day. Flies caught in such rounds are recorded as to sex and hunger stages, then marked and released, with the exception of those engorged. Resting flies are also searched for in order to obtain additional specimens replete with blood. A note is also made about the frequency with which game are observed during fly rounds. These fly rounds would provide the usual data about monthly densities, sex ratios, recaptures, etc. But they would also provide information about (*a*) fly activity at sampling stations at different times of the day; (*b*) information on any changes in feeding habit that occur, and (*c*) a comparison of the feeding habits in relation to the three different types of terrain and habitat traversed by the fly rounds.

Another example of additional information which can be provided by introducing slight modifications in the basic fly-round technique comes from Northern Nigeria (Jordan, 1965). In catches of *Glossina morsitans submorsitans* made in the course of the fly round, separate records were made of the numbers of flies caught on the body of one of the catching team, and the number taken on vegetation and on the ground in the vicinity of the team. The results (Table I) show a wide difference in the

TABLE I

Number of *Glossina morsitans submorsitans* caught on bodies of catchers, and on the vegetation or ground (after Jordan, 1965).

	Total Catch	Males	Females	Females as % of total
Body	2 378	1 534	844	35·8%
Vegetation or ground	5 034	4 750	284	5·6%

partition of the two sexes between the two sites. In the case of the males, 75% of the total number were taken on vegetation and on the ground, as compared with 25% on the body. Of the total females, the proportions were reversed with 25% on vegetation and ground, and 75% on the catcher. If these figures are expressed in another way it is seen that of the total tse-tse population taken, the females formed 36% of those taken on the body, but less than 6% of those taken on vegetation and ground.

The fly round has also been developed in Nigeria, West Africa and adapted to deal with sampling of the riverine tse-tse — especially *G. palpalis* and *G. tachinoides* — which are closely associated with man. The method long used in Nigeria is to lay out a linear fly round to sample *G. palpalis* along a selected part of a stream. The sections are subdivided into 100-yd long subsections and — in order to prevent a catching-out effect — all flies are recorded, marked and released. These fly rounds were carried out along a fixed route at regular intervals over a period of six years in a study of apparent density, longevity, and movement of marked flies in relation to season and vegetation, etc. (Nash and Page, 1953). A similar technique was used later for studying the effect of obstructive clearing directed against *G. palpalis* (Nash and Steiner, 1957). This riverine fly round is considered particularly efficient because it samples the very area to which the riverine tse-tse are linearly confined. In contrast to the experience with the savannah species of East Africa is the fact that allowing for certain seasonal variations, this riverine fly round samples males and females in approximately equal proportions.

The fly round technique has also been used in Nigeria to study *G. longipalpis*, a species which does not feed readily on man (Page, 1959a). Initially a 3 200-ft track, divided into six sections and subdivided into 100-yd subsections, was used. But later, due to the limited period of daily fly activity, only certain representative subsections were sampled. As this species has a pronounced daily rhythm of activity (being inactive below about 23°C) fly rounds were not started until that temperature was reached. It is interesting to note that with this species the bulk of the flies taken in the fly round were caught on vegetation rather than in the vicinity of the bait. The effect of type of bait on the sex ratio of flies was also a striking feature; with man bait, females formed only 1·9% of the total, while with ox bait, they formed 39%.

2. *Stationary bait*

While the fly round, based on collections carried out in a three or four mile transect of different vegetation types, has proved of considerable practical use with the game tse-tse, such as *G. morsitans* which are widely dispersed through great areas of savannah woodland and with the riverine tse-tse in certain areas such as Nigeria, experience in other parts of West Africa such as Ghana has shown that the species limited to riverine forests (*G. palpalis*) or to forest clumps or thickets (*G. tachinoides*) can be effectively sampled by a stationary team confined to a well-defined habitat, such as a river crossing (Morris, 1961a, b).

With the stationary team it is found more feasible to extend the period of catching to eight hours, or even the whole day (Page, 1959a, b, c). In this way any variables due to daily fluctuations in the activity cycle will be automatically dealt with. The extended period of observation also serves an additional function — sometimes the main function — of being able to record tse-tse at very low densities. With *G. palpalis* in particular — the main vector of human *Trypanosoma gambiense* — it is frequently found that tse-tse densities may be extremely low even in areas where there is a high incidence of human trypanosomiasis (Nash, 1952, 1960). Under those conditions it may require a whole day's catch to detect the occasional fly.

The stationary-bait method has been used in East Africa in studies on *G. pallidipes* in connection with transmission of animal trypanosomiasis (Leggate and Pilson, 1961; Pilson and Leggate, 1962a; Harley, 1965; Pilson and Pilson, 1967). Continuous catching of tse-tse is carried out on a tethered black ox from before dawn until after dusk. The advantage of this method is that, firstly, it can deal adequately with tse-tse which show a marked peak of biting activity at certain periods of the day — for example, *G. pallidipes* has a large evening peak of biting. Secondly, it catches both males and females. Thirdly, by allowing flies to actually feed on the ox before they are captured, the capture can be interpreted strictly in terms of hungry flies. On this basis it is possible to record the observation in terms of "mean number of non-teneral flies feeding on ox per day". In view of the fact that some species of tse-tse are liable to be taken on bait at any hour of the day or night (Robertson, 1962) the logical development of the 12-hour day catch is to extend it to the full diel or 24-hour period. This refinement has now been applied to three species of tse-tse in Uganda known to differ in activity cycle, and a careful hour-to-hour record has been kept of both males and females of these species, with regard to different seasons of the year, as well as with reference to whether the bait animal was sited in the open or in the shade (Harley, 1965).

A great deal of new information has been obtained in this way which has an important bearing on the relation between sampling procedures and the different activity patterns of the tse-tse flies concerned. For the present purpose it would be sufficient to illustrate this by means of the particular figures referring to the catch of females taken on bait in open sites in the dry season (Fig. 1). From this it is seen that *Glossina pallidipes* shows a steady increase in biting activity throughout the course of the day, with a peak in the late afternoon. *G. palpalis fuscipes* shows low activity in the early morning and late afternoon, with peak activity in the middle period of the day. *G. brevipalpis* has very low biting activity throughout the day, but exhibits two sharp peaks at dawn and at dusk, immediately before sunrise and immediately after sunset. Slight differences exist in the actual form of the curve between males and females and there are also slight differences in pattern according to the season of the year, and whether the bait is sited in the open or in a shady position.

One feels that in the general context of vector sampling it would be an additional advantage to have some idea of the effective attraction radius of a single, isolated host animal. Possibly, synchronized collecting on a number of isolated oxen tethered at equal intervals — which could be varied from experiment to experiment — throughout blocks of bush might clarify this point, and might even be used as a method of estimating the total tse-tse population in an experimental block. The variable attraction of different bait animals may also have to be taken into account. In Northern Nigeria, for example, it was found that a white Zebu bull proved to be the best bait for one particular species of tse-tse. Experience elsewhere has shown that for another species, *G. pallidipes*, black cattle attract significantly more tse-tse than white and brown ones, although no preference between brown and white was observed. In still another series of observations involving two black Zebu oxen of similar size, it was noted that one of the bait animals consistently attracted larger numbers of all the species of tse-tse caught (Saunders, 1964). Presumably, individual attraction of bait animals may be liable to vary for other reasons, size, odour, etc. The general attraction to the tethered bait may also be influenced by the presence of alternative hosts in the neighbourhood, or by the increased tse-tse activity associated with high tse-tse populations or with the presence of big game (Power, 1964). For predominately day-time biting flies like tse-tse, one might also expect that seasonal changes in the vegetation, e.g. prolific growth of long grass in the wet season, might limit the range of visual attraction.

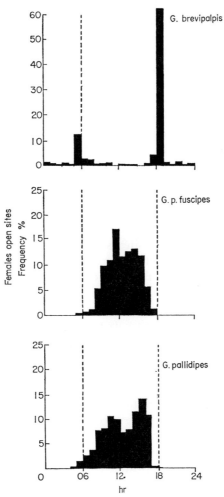

FIG. 1. Activity pattern of three species of tse-tse, *Glossina brevipalpis*, *G. p. fuscipes* and *G. pallidipes* as exemplified by records of females sampled in open sites. (Harley, 1965).

3. *Trapping*

A variety of traps have long been used for tse-tse studies, not only as additional sampling methods, but as a basis for control when adopted on a sufficiently large scale. The design and use of these traps has varied somewhat according to the investigator, the ecology of the species studied, and the particular use for which the trap is designed. Rather than deal with the range employed, it would be more instructive from the point of view of sampling in general to describe one particular design which has been the subject of special study, not only in West Africa where it originated, but also in East Africa where its use has been extended to a different range of tse-tse species. This type has been designated the "animal trap" (Morris and Morris, 1949; Morris, 1960, 1961a, b, 1962) and is designed to resemble a small host animal about the size of a goat. The most important features are the cylindrical shape of the body, producing highlights and shadows. This curved body of hessian or cloth is open at the bottom. A slit along the upper surface has a cage superimposed into which the flies are attracted — through a non-return slit — from under the body. When the use of this trap was extended to East Africa it was found necessary to use a larger design (4 ft long, 2 ft deep and 4 ft high) to attract the species there. In utilizing these traps to the best effect considerable emphasis is laid on appropriate siting with regard to open feeding grounds of tse-tse, such as ferries, water-holes, etc. The attributes of a "good" site have never really been elicited, and in many instances the exact location of the most productive sampling or trapping sites is arrived at simply by trial and error according to local circumstances. Other requisites are the regular presence of natural hosts in the locality, good visibility of the trap from a wide angle, and arrangement of traps at right angles to lines of movements of flies.

Originally the animal trap was used in studies on *G. palpalis* and *G. tachinoides*. With *G. palpalis* this design of traps appears to be much more effective than several other accepted designs (for example, twenty times more than the Harris trap, ten times the Swynnerton trap, and seven times the Chorley trap). When used with care and discrimination it appears that this trap has certain advantages in sampling, particularly in primitive countries where the variable human factors may influence other catching methods. Traps can also be used simultaneously over a wide area to indicate preferred habitats, and as they can be left *in situ* for long periods they are less affected than other methods by density fluctuations throughout the day, or from day to day.

It is claimed that the efficiency of these traps is self-evident when they are used on a mass scale, as they may collect such a high sample of the population as to produce virtual eradication in an area. In such a case the trap figures recorded could approximate closely to a true population count. For routine sampling this feature of the technique could be a complicating factor in that, if a large number of traps are used, the figures may indicate a sharp initial fall in the population which is followed by a more uniform rate of catching. This is an interesting situation in which the sampling method is also functioning as a control measure (Morris, 1961a, b, 1962). Opinions among tse-tse workers themselves about the general value of this method of sampling tend on occasions to be conflicting and controversial. Perhaps the most objective way of evaluating this technique is to examine the results of carefully controlled tests in which trapping is compared with one or more alternative sampling methods in the same locality. Observations of this kind are described later in the chapter.

Before continuing with the review of other sampling methods used in tse-tse studies it would be useful at this stage to point out that the potentialities of artificial animal-like traps have also been explored — quite independently in many cases — with other groups of day-time biting flies, particularly horse-flies and deer-flies (*Tabanidae*) and black-flies (*Simuliidae*). In connection with the former group of insects, studies in Canada have been particularly instructive, especially those concerned with the development of the "Manitoba fly-trap". (Bracken *et al.* 1962; Thorsteinson *et al.* 1965). This work originated in the observation that horse-flies were attracted to a weather balloon, and a balloon of this type was tested by inflating it to 24 in. diameter coating it with tanglefoot, and suspending it 4 ft above the ground. Later, spheres which could be coloured, were suspended as a decoy below a fly-trap with a translucent canopy. Later still, the rubber balloons were replaced by polystyrene spheres inflated to 20 in. diameter and the non-return trap was charged with sodium cyanide in order to kill the flies immediately on entry. In its final stage, the Manitoba fly-trap consisted of a spherical decoy target made from two hemispheres constructed of black acrylic plastic sheets. The trap chamber is capacious enough to hold approximately 2 000 horse-flies, and the trap performance can be judged by the fact that in some areas capture rates of the order of 1 000 female Tabanidae per hour were recorded. Trap capture data have formed the basis of information about seasonal and geographical distribution of Tabanidae in Manitoba, as well as essential data on species composition (Hanec and Bracken, 1964).

Behaviour studies aimed at analysing the attraction of these silhouette traps to horse-flies (Bracken and Thorsteinson, 1965), have examined such factors as colour, shape, movement of target, trapping site and so on, all of which features demand equal consideration in any attempt at analysing the attraction of tse-tse flies to "Animal traps".

Perhaps even closer to the problem of sampling tse-tse by means of "animal traps", are the parallel studies on "animal or silhouette traps" for sampling day-time biting *Simulium* in North America (Fredeen, 1961). These will be discussed in more detail in a later chapter (page 99) but suffice to say at the moment that these traps constructed in the form of a 4-legged animal (cow silhouette and sheep silhouette) appear to be very close in principle and design to the tse-tse "animal traps" discussed previously. Again, analysis of the various factors determining the attraction of these silhouettes to *Simulium*, together with allied work in Germany on the attraction of different species of *Simulium* to specific parts of animal and bird silhouettes (Wenk, and Schlorer, 1963) may prove of direct value in appraising the role of this trapping or sampling method in tse-tse work.

4. *The resting population*

For many years tse-tse sampling methods were mainly concerned with the active fly caught in the course of the fly round, or coming to traps and animal bait. Studies on the resting or inactive fraction of the population only received due attention at a much later stage (Nash and Davey, 1950), probably because the search for resting tse-tse — as distinct from active flies which have settled temporarily — appeared laborious and unrewarding. The recent increased study of resting populations perhaps owes its stimulus to the following causes:

    i The realization that as females of most savannah tse-tse form such a small proportion of the accessible tse-tse population (as judged by fly-round catches) they must obviously be resting somewhere in large numbers, in accordance with the estimate that there are more females than males in the population (Isherwood, 1957).

    ii The need in survey work to detect the presence of tse-tse species which do not come readily to human bait, and which consequently may be overlooked although they are epidemiologically important (Nash and Davey, 1950; Nash, 1952; Page, 1959c; van den Berghe and Lambrecht, 1954).

    iii Intensification of the search for resting engorged flies — as distinct from those containing digesting blood taken in the fly round — in order to provide better samples for precipitin studies on the blood preferences of different species of tse-tse (Weitz and Glasgow, 1956; Jordan *et al.* 1960, 1961).

    iv The need for more precise knowledge of tse-tse resting sites according to location and vegetation in connection with the efficacy of insecticide application from the air or from the ground (McDonald, 1960a, b; Kernaghan, 1961; Chadwick *et al.* 1965).

Day-time studies on the resting population of *G. swynnertoni* (Isherwood, 1957) disclosed, as expected, a much higher proportion of females than in the active catch. These studies also showed that in the search for the day-time resting population the results are liable to be affected by the fact that active flies follow the collectors and settle on the trees. This variable was dealt with as far as possible by collecting active flies first.

A most interesting further development in the sampling of the resting population has been the investigation of nocturnal resting places. In West Africa resting females of *G. palpalis* proved very elusive but were observed by day on twigs and leaf petioles about one metre above the ground where they would remain motionless for several hours (McDonald 1960). By using fluorescent paint as a marking agent, with subsequent detection by ultra-violet light (see also Jewell, 1956), the night-time resting sites could be investigated for the first time. It was shown that resting rarely took place at a higher level than 2 m above the ground, half the liberated flies resting on leaves and small twigs (never on tree-trunks) within 30 cm of the ground.

Related observations in East Africa, using as a marking agent microscopic glass beads such as those used in road reflecting paint, have not only literally "thrown light" on the nature of the night-time resting places of *G. swynnertoni*, but have also shown that after dusk resting tse-tse move to a different type of resting site from that occupied by day (Jewell, 1958; Rennison *et al.* 1958; Southon, 1958). A modification of the same technique has been used to study the resting sites of *Glossina morsitans* in Zambia (Robinson, 1965). Captured flies were immobilized and marked with a reflecting patch (white Scotchlite) easily visible at night at a distance of 20–30 yds by the beam of a torch light. Marked and released flies had the chance of feeding on a tethered calf and they were thus encouraged to rest near the release point. By means of a graduated pole it was found that (22%) a considerable number of *morsitans* were resting above 4 m and that the highest level was 20 ft. It is worth noting that information of this kind, dealing with the exact resting sites, had a direct bearing on the planning of a campaign against this species based on spraying routine.

A further development of the technique of sampling the resting population of tse-tse has been based on the use of the marking-release-recapture method (Chapter 8). The object of these studies carried out in Southern Rhodesia (Pilson and Leggate, 1962b) was to find inactive *Glossina pallidipes* in given vegetation types, and also find if resting behaviour varied with season and/or time of day.

Flies caught on a tethered ox were allowed to feed, and were then marked on the 25000 system (see Chapter 8), all details of sex and physiological condition being recorded. The marked flies were then released, and a subsequent search for resting tse-tse was carried out in the vicinity of the ox over an area of about one acre on all possible resting sites, including the upper branches of some trees up to a height of about

45 ft. Under those conditions the recovery rate was high: 10–30% of marked males, and 10–17% of marked females. Eighty-nine per cent of the total resting flies were taken at the 0–9 ft level. The catch also revealed that there were seasonal variations in resting behaviour, the percentage seen at 3 ft and under being much greater in the late hot dry season. In the rainy season and cool season, flies were found mostly on branches.

In a follow-up of these observations (Ford, 1962) some precise figures were obtained about the change in preferred sites at different times of the day. The figures for *Glossina pallidipes* are shown in Table II, and

TABLE II

Distribution of *Glossina pallidipes* on boles, branches and rot holes in the Zambesi valley in the hot season of October (after Ford, 1962).

Time of observation	Boles	Branches	Rot holes
0900–10 h	10	13	41
1000–11 h	17	2	54
1100–12 h	29	1	112
1200–13 h	29	—	152
1300–14 h	41	—	?
1400–15 h	43	—	75
1500–16 h	23	—	87
1600–17 h	18	—	12
1700–1745 h	4	5	1

demonstrate that during the heat of the day branch sites are evacuated and rot holes become attractive. In addition, there is a general move to lower sites, 3 ft or less. When the same basic methods were later used to study the resting habits of *G. morsitans* (Pilson and Pilson, 1967), the search for resting flies was initially confined to 4 transects, each 20 ft wide and 200 ft long, radiating out from the stationary ox in four directions. It was hoped that this design would show whether tse-tse which had fed on ox would rest near, or well away from it. The search for tse-tse was aided by 6 ft and 25 ft ladders, and it was found that the tse-tse could be detected with ease at height ranges from 4 to 40 ft. However, the impossibility of searching upper branches of trees prevented a composite picture of overall diurnal resting being obtained.

In the context of vector sampling in general, one of these tse-tse investigations on resting populations (Chadwick, 1964) has been particularly illuminating in that it attempts to answer a basic question, viz. what proportion does the resting population form of the total vector population in an area? In Tanzania a wide range of potential resting sites were examined, and it was found that for the tse-tse in question — *G. swynnertoni* — 65% of resting flies were found on the undersides of branches. An attempt was then made to relate the number of flies found on branches to the total number of flies expected to be in the area searched. Two fly boys carried out a normal fly round (see page 9) while three searchers examined a strip 40 yd wide, for some 20 sectors, each 100 yd long. The "apparent density" was worked out using a standard availability of 10% and the observation that there are twice as many females as males in any given area. The figures are shown in Table III.

TABLE III

Relation between number of tse-tse flies (*Glossina swynnertoni*) recorded on branches, and the total number of flies expected to be in the area searched (after Chadwick, 1964).

Trial	Number of flies found resting on branches	Apparent density (from fly-round)	Number of flies expected†	% found
1	100	307	238	42%
2	33	122	85	39%
3	45	174	128	35%

† In 40 yd wide strip

Subject to error, the results are interpreted as suggesting broadly that in the area in question about one third of the total tse-tse fly population could be found resting on the undersides of branches.

There are other vectors such as anopheline mosquitoes about which a great deal is now known concerning outdoor resting captures as compared with indoor collections and bait collections. But it has not yet been possible to be anything more than highly speculative when trying to estimate what proportion of the total mosquito population in an area is represented by any one of these samples.

COMPARISONS BETWEEN TRAPPING AND OTHER METHODS OF SAMPLING TSE-TSE

It is often difficult to appreciate the advantages or limitations of a particular sampling method in absolute terms. A simultaneous comparison with other sampling methods can however provide the first step towards sounder evaluation. Brief reference has already been made to the comparative results obtained with the animal trap in competition with other trap designs. More instructive, however, is a comparison between traps and the conventional fly-round or fly-boy catch. There are illuminating figures on this point from both West Africa, mainly *G. tachinoides*, and East Africa, *G. pallidipes*.

In West Africa a comparison was made through a complete year between trap samples (150 trap days) and fly rounds, 20 fly-boy days monthly, working simultaneously (Morris and Morris, 1949). These figures showed that the relative efficiency of the two methods differed at different seasons of the year (Fig. 2). In January, the dry season,

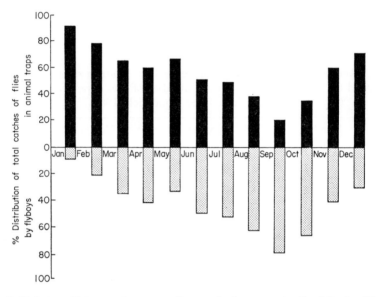

F I G. 2. Relative efficiency of two sampling methods — standardized "animal" trap catch and fly-boy catch — for *Glossina tachinoides* at different seasons of the year in Gold Coast (Ghana) (after Morris and Morris, 1949).

during the period of the cool, dry "harmattan" the traps were ten times more efficient than the fly boys. At the other extreme, in September at the end of the rainy season, the traps caught only one quarter of the fly-boy catch.

A similar comparison was made between "animal traps" and fly-boy catch in East Africa (*G. pallidipes*) (Morris, 1960). Over a period of six weeks, eight traps which were easily supervised by one boy clearing them five times a day, were compared with a fly-boy team of two, making morning and afternoon rounds. The results were as follows, at a time of high fly density:

Traps: 1586 flies with 82% female
Boys: 165 flies with 13% female

At lower densities, in a five week trial, the results were:

21 traps visited daily: 30 flies with 73% female
2 boys, 2 rounds per day: 9 flies with 30% female

These figures bring out clearly not only the high gross catch of the traps, but also the much higher proportion of females taken. In fact, the proportion of females is very close to what is estimated to exist in nature. (The two sexes of tse-tse are known to emerge from puparia in approximately equal numbers, but as females have a considerably longer life span than males, a wild population of tse-tse is reckoned to contain 70%–80% of females (Jackson, 1949).) This consistently high proportion of females taken in traps has been encountered in all species of tse-tse tested, indicating that from that point of view the sample appears to be more representative of the whole population. It should be pointed out that in this experiment with *G. pallidipes*, the marked superiority of the "animal trap" is a little misleading in that it is being compared with a single alternative sampling method which has long been known to be rather unproductive of this species. *G. pallidipes*, unlike *G. morsitans*, is not greatly attracted to human bait in the normal fly round, much preferring oxen or cattle whether moving or stationary. With this in mind the following experiment, involving alternative animal bait, is perhaps a more realistic comparison.

An extended comparison using three different methods of tse-tse sampling — "animal trap", cattle bait and fly round — has been carried out in Uganda (Smith and Rennison, 1961). The experiment was based on the 8×8 Latin square design using oxen of different colours and traps of two colours, black and brown. The fly round was carried out on two patterns:

    i. two or three times weekly, from 08.00 hours to 09.30 hours, along a path 3000 yd long divided into 50-yd sections,

    ii. 1000 yd portion of fly round, along which experimental methods of catching were used, was traversed seven times daily in each direction between 08.00 hours and 18.30 hours.

The results of the first series of observations are shown in Table IV.

This shows that more flies were taken on cattle than in traps, but the traps recorded a higher proportion of females. On some days trap catches were almost nil at a time when cattle continued to attract numerous flies. In addition, the period of the day when most flies were taken differed in traps and on cattle; with the cattle most flies were taken in

TABLE IV

Numbers and sex ratio of non-teneral *Glossina pallidipes* caught by three different sampling methods used simultaneously, Morris ("animal") traps, bait oxen, and fly round (after Smith and Rennison, 1961).

	Method of capture	Number of flies	
		Males	Females
Expt 1.	Brown trap	1341	3916
	Black trap	884	2013
	Oxen	8344	5292
	Fly round	213	123
Expt 2.	Brown trap	213	504
	Oxen	3011	1774
	Fly round	604	97
	Fly round	299	159

the morning and the evening. In the traps most flies were taken between 12.20 and 14.00 hours, and least in the morning. From these and other data it appears that the traps are sampling a different proportion of the population from that attracted to the cattle. The traps, despite their animal-like appearance and design, are not simply functioning as substitute annimals; they must be providing an additional attraction for those flies, especially females, seeking a shaded resting place in the hotter, brighter periods of the day. Confirmation of these findings is provided by observations in the rain forest area of Liberia, West Africa (Foster, 1964) which indicate that the theoretical basis for the attraction of "animal traps" is still very doubtful. Although the trap is designed to resemble a host animal, the live animal itself, e.g. goat, does not attract *G. palpalis* in that area.

From the fly-round catch it was shown that the number of females taken followed the same general trend as females caught on the oxen, suggesting that females taken in the fly round are mainly hungry flies. This experiment also showed that, while the fly-round catch was fairly uniform in the wet season and in the dry season, both oxen and trap catches were twice as high in the rains as in the dry season.

A simultaneous evaluation of four different methods of sampling tsetse has been carried out on *Glossina swynnertoni* with reference to the hunger cycle of the non-teneral males (Bursell, 1961). The methods used were:

i. flies caught approaching or attacking a catching party (author and two African assistants). This was referred to as the "Standard" catch;

ii. flies caught attempting to probe skin of young heifer, referred to as the "Bait" catch;

iii. flies caught on perching places on trees, the "Resting" catch;

iv. flies attracted to slow-moving lorries, the "Vehicle" catch.

The hunger cycle of the male tse-tse starts with the engorged fly, and is marked by a high rate of digestion of the blood meal, the products of digestion being converted into fat. Partial digestion follows, and the culminating phase is hunger and fat depletion. An analysis of the catches taken by these different sampling methods showed, not altogether unexpectedly, that the different stages of the hunger cycle were not equally represented in each catching method. While the gorged flies were unresponsive to the stimuli of moving objects and to bait animals, the reaction of the later stages of the hunger cycle were less predictable. While activity in general increases, the penultimate stage of the hunger cycle is well represented in the resting catch, and has also a lowered attraction to moving objects.

A considerable advance in our knowledge about the relative validity of different sampling methods was made possible by developments in age-grouping techniques (Saunders, 1960, 1962). By means of a careful study of the ovarian development of female tse-tse it became possible to grade females up to the fifth cycle of ovulation, i.e. up to the time the insect is about 50 days old, but not beyond.

Age grouping by this method was carried out on G. pallidipes populations sampled by four different methods:

i. modified Morris "animal" traps;

ii. by hand in standard fly round (author and two assistants);

iii. on bait animal (Zebu ox tethered in shade);

iv. resting flies on branches in thickets.

These investigations showed that female G. pallidipes entering Morris traps are older than those caught in the fly round, while those caught on bait animal are intermediate in age structure. Correlated with age structure is an increase in the proportions of females carrying third-instar larvae. In general, therefore, the oldest females and those most advanced in pregnancy are more readily taken in traps, while the youngest and least advanced in pregnancy are more liable to be taken in the fly round.

Female tse-tse captured in these experiments could be classified into five groups (O, I, II, III and IV) according to the numbers of ovulations completed. This still meant however that the last group — IV — was much less clearly defined than the others as it included all flies which had ovulated four or more times, and therefore embraced a very wide range of older flies. Later refinements in this technique worked out with other species of tse-tse (Challier, 1965; Harley, 1967a, b) have now made it possible to extend this physiological age-grading to eight groups (O–VII) according to the number of ovulations, thus enabling a more precise subdivision of the older fraction of the female population — up to about 80 days old. Tse-tse workers point out however that even this noteworthy advance in technique still does not extend far enough to give a reasonable picture of the age composition of a wild population of female tse-tse.

These advances in physiological age-grading of female tse-tse flies have opened up new possibilities of investigating the real or calendar age (in days) of wild caught females, and have also renewed the possibility that this work in turn might throw light on the equally important problem of the age composition of the male tse-tse population. As the techniques used to establish the relationship between physiological age and real or calendar age have been based on the marking-release-recapture method, this aspect will be discussed more fittingly in Chapter 8 along with similar problems encountered in other fields of vector ecology. The application of these new age-grading techniques to sampling problems has been well brought out in a very comprehensive comparison of no less than seven different sampling methods carried out simultaneously on *Glossina pallidipes* and *G. fuscipes* in the Lake Victoria region of Uganda (Harley, 1967a, b). The methods used were as follows,

    i. Flies caught on a stationary black ox attended by two catchers.
    ii. Flies caught on a moving (walking) black ox attended by two catchers.
    iii. Flies caught on a moving vehicle (driven at walking pace) attended by two catchers.
    iv. Flies caught in a Morris trap and a modified version of a Harris trap placed three yards apart and from which flies were removed hourly.
    v. Flies caught while at rest on bushes and trees, the flies following the 4–6 searchers having been caught and killed as far as possible.
    vi. Flies caught on two stationary men.
    vii. Flies caught on two moving (walking) men.

In the first of two series of experiments the collections on stationary and moving parties took place in the same locality, the moving party walking, or driving, round the circumference of a circle about 12 yd in diameter, and the stationary catch being made at the centre of the circle. Seven catching stations were used, each catching method being used for one day at each of the stations in a randomized manner. In a second series, the moving catches were carried out along tracks through the bush, and therefore sampled completely different and much larger areas compared with the stationary catch.

Females of the two species of tse-tse captured in the first series of experiments were divided into the eight physiological age grades, O–VII, described above, thus providing information on the age composition of the different samples in addition to the information already provided about relative numbers caught by different methods, and relative proportions of males and females.

In accordance with the main theme of this particular investigation, males and females of the different samples were also dissected for trypanosome infection, and consequently the experiments provided an impressive mass of information about many different aspects in the ecology of these two tse-tse flies. For present purposes, however, the application of the age-grading technique to sampling problems is the most noteworthy feature of this comprehensive experiment. Differences in age-composition between the different samples were much more marked in the case of *G. pallidipes* than *G. fuscipes*, and the former has therefore been selected to best illustrate the extremes likely to be encountered (Fig. 3). Making allowance for the fact that the samples of resting flies and flies caught on man are not strictly comparable with the other samples (as low catches in the first series made it necessary to include additional data from further catches) the findings do bring out the striking difference in age-composition between samples caught in traps and on oxen, and those caught on a moving vehicle, on man, or while resting.

THE EFFECT OF ADVERSE CONDITIONS—NATURAL OR MAN-MADE—ON THE VALIDITY OF TSE-TSE SAMPLING

Some very stimulating information on this aspect of sampling, which is one of direct concern in many other fields of vector control, has been provided from three different sources:

1. Evaluation of aircraft spraying with insecticide as a method of tse-tse control (Foster *et al.* 1961; Burnett *et al.* 1961, 1965).
2. Studies on outbreaks of human trypanosomiasis in northern Ghana (Scott, 1960; La Croix, 1960);
3. Studies on the extermination of tse-tse, *G. morsitans*, by fire exclusion and by discriminative clearing in Northern Rhodesia (Glover *et al.* 1955).

1. *Insecticide spraying by aircraft*

Some very illuminating features have emerged from recent work on the aircraft application of insecticide against the tse-tse of the savannah

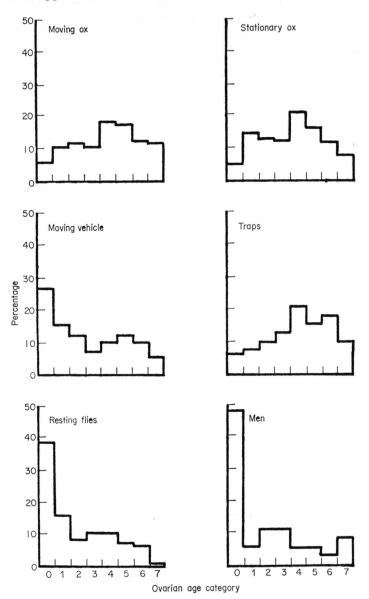

FIG. 3. Age composition of samples of female *Glossina pallidipes* collected by different methods (first series) (after Harley, 1967).

woodland of East Africa (Foster *et al.* 1961; Burnett *et al.* 1961; Burnett, 1961a, b, 1962). Evaluation of this measure was based mainly on the apparent density, i.e. the number of non-teneral males caught per 10000 yd of the fly path. In addition to this, however, the catch recorded the number of teneral males in order to indicate the extent to which flies continue to emerge, after insecticide application, from pupae which are unaffected by insecticide. The number of females, both teneral and non-teneral, was also recorded. Search was made once or twice a week on seven fly rounds in the experimental block. In the fly round human bait was used for *G. morsitans*, and either cow bait or hessian cloth on wooden screens for *G. pallidipes*. The interpretation of these catches is made difficult by two important factors. Female flies, which are in general longer-lived than the males, only come to man mainly in the first three weeks of life; the older reproductive fraction of tse-tse is therefore not readily available by fly round and other methods employing human bait. In addition, older females are also less susceptible to insecticide than younger females, and may persist despite a "high kill" of younger females, and males, which are mostly young and susceptible. As a result, the very fraction of the population liable to survive and reproduce is the one which is least likely to be recorded by the sampling techniques used.

2. *Outbreaks of human trypanosomiasis*

Of the two vectors of human trypanosomiasis in northern Ghana, *G. palpalis* is the one of major importance, and the one which is the principal, if not the sole vector responsible for high incidence and outbreaks of this disease. In some of these outbreak areas *G. palpalis* is approaching the limits of its natural distribution, climatic conditions restricting the fly to a narrow riverine habitat. It is under these very conditions that man-fly contact will be at its closest and most dangerous. The epidemiological implications of this are of considerable significance, but for the purpose of the present review the main lesson is that sampling based on the use of "fly-boys" or human bait under those conditions will vary considerably in its efficiency or validity according to the size and degree of aggregation of the tse-tse population. Tse-tse occurring at very low densities in unfavourable areas may be much more readily available for capture simply because they are much more concentrated into strictly limited localities than the larger populations which occur wherever suitable habitats are widespread and tse-tse are not confined to a particular concentration site. Similar difficulties in interpreting sampling data at climatically unfavourable times of the year have been disclosed in more recent observations in Rhodesia (Bursell, 1966). In Rhodesia it is generally found that during the hot dry season fly-round catches are very much higher near riverine vegetation than elsewhere. This has led to the belief that tse-tse populations at this time of the year are dependent on riverine vegetation for survival. However, by employing two types of sampling (fly-round catch, and catch on stationary ox bait) in four different vegetation types, it appears that such differences in fly-round catch show only that the sexually appetitive males are concentrated on riverine vegetation, and that other elements of the tse-tse population may be quite evenly distributed between different vegetation types.

3. *Fire exclusion and discriminative clearing*

In a control scheme based on the densification of woodland by means of fire exclusion and on discriminative clearing, the fly round showed an apparent density of about 50 before control started. In the following years after the first clearing the figures were 15, 1·7, 0·1 and finally 0·0. In 1942 when fly numbers had already been greatly reduced, an uninseminated, but non-teneral female was found (normally, all females which have taken their first blood meal are inseminated). In the following year 9% of non-teneral females were uninseminated, 20% in the next year, and 29% in the subsequent year. Allowing for the fact that at these very low apparent densities the exact figures may become a little unrealistic, this example provides still another factor for consideration when dealing with sampling of very low vector populations and with the evaluation of suspected "vector eradication". When a vector population is drastically reduced by adverse conditions or by applied control methods, there must be some point at which the rapidly decreasing chances of contact between male and female may become a limiting factor. Very little is known about this critical level, or its possible range of variation with different vector insects. The subject is clearly one requiring further attention in the general context of sampling at very low vector densities and in the interpretation of low-density data.

CHAPTER 4

ANOPHELINE MOSQUITO VECTORS OF MALARIA

Of all insect vectors of disease, anopheline mosquitoes have received the most universal attention over the longest period of time. As a result there is a vast literature dealing with every country in which malaria has occurred, and with the wide range of species responsible for transmitting that disease.

In many countries of the Old World in which so many of the original classical studies on *Anopheles* were carried out, the close association between the main vectors and both man and his habitations was early emphasized. Many of the principal vectors of malaria in Europe, Asia and Africa were recognized as being "domestic" mosquitoes in that the female mosquito not only entered houses to feed in large numbers, but also remained in these houses or habitations during the day, provided that suitable dark, day-time resting places were available. It was early recognized that there were exceptions to this general concept, and workers in the Philippines, in Malaya and in Central and South-American countries found that, while their particular vectors would readily enter houses at night in order to obtain blood meals, they seldom remained there in appreciable numbers by day. However, in general the house or stable formed the main centre of catching and sampling activities; and this house-catch, usually by day, provided the great bulk of the information about distribution of vectors, seasonal abundance, effect of malaria control by antilarval methods, etc. In all these catches the main object of attention was the blood-feeding female mosquito. The male mosquito usually received only incidental attention when it was taken resting indoors along with the females. Occasionally, a high proportion of males in the day-catch was used to indicate proximity of breeding grounds; in the same way, persistence of males in the catch was also used as an indication of continued emergence from breeding grounds.

A further impetus to the study of house-resting or house-haunting mosquito populations was provided by the work carried out prior to and in the first years of World War II on the control of anophelines by regular house spraying (space spraying) with pyrethrum in kerosene. The idea of the house as the main centre of attack against adult mosquito vectors of malaria developed still further with the discovery and application of the residual insecticides. These developments provided further impetus to the use of the house or habitation as a sampling centre, often to the exclusion of other methods or other non-domestic collecting sites. As a result of this preoccupation it was too readily assumed that the house-resting catch was a true reflection of the vector population as a whole, and some of the remarkably drastic results claimed in the early years of DDT application in malaria control campaigns were based on the almost complete disappearance of the day-time resting population from treated houses.

Within the last few years the increase in development of the global malaria eradication programme has demanded a new and more critical approach to the question of evaluation and to the question of the validity of various catching and sampling methods. This new critical self-appraisal, very similar in trends to that described in the study of the tse-tse populations, has been encouraged by the increasing facility for exchange of information and by the increasing opportunities for workers from different countries to meet those from other countries where conditions and experience may be totally different.

In this present review no attempt will be made to do full justice to the vast and expanding literature on the ecology and sampling of anopheline mosquitoes. Only a general outline will be possible in which references will be made to a comparatively limited number of studies which have been selected because of their bearing on the general thesis of vector populations as a whole. Full details of a wide range of capture and sampling techniques, with special reference to the entomology of malaria eradication, can be found in a manual produced by the World Health Organization (W.H.O. 1963).

The sampling methods at present in use in anopheline ecology can most conveniently be discussed under the five headings below. In view of the fact that there are several sampling methods which could well be discussed under more than one of these headings, these headings should not be considered to be sharply-defined categories. For example, the bed-net technique and its modifications, long used by the Malayan group of workers for trapping mosquitoes which enter houses to feed at night, but which do not remain indoors by day, could be equally well discussed under headings 1, 3, and 4 below. As it is, the use of this particular technique indoors is examined in Chapter 5 on culicines, and its application to outdoor biting anophelines is discussed under heading 4 below.

1. Catch of day-time resting populations indoors, supplemented on occasions by night-time collections of resting mosquitoes.

2. Catches of the day-time resting population in outdoor resting places.
3. Catches of female mosquitoes biting human and animal bait indoors and outdoors.
4. Trapping methods for the automatic capture of mosquitoes coming to bait or attempting to leave houses and habitations after feeding.
5. Combinations of different methods.

All these methods are primarily concerned with the catch of female mosquitoes and are all closely related to the cyclical rhythm of feeding, blood digestion and development of ovaries, and oviposition followed by further blood feeding, known as the gonotrophic cycle. While a few "highly domestic" vectors may spend the greater part of this cycle (normally two to three days in tropical countries) indoors; others may feed indoors but leave for outdoor resting places half-way through the cycle or after an even shorter interval. The different degrees of behaviour in this respect, shown by different species of *Anopheles*, play an important part in their availability by different sampling methods. As this behaviour in relation to feeding and resting is liable to be considerably affected by the mass spraying of houses with residual insecticides in malaria eradication programmes, it follows that the relative significance of the different concentration sites, which form the main foci of sampling activities, may likewise be drastically altered. An appreciation of this point is not only essential for evaluating the effect of insecticide pressure on vector populations, but it is also essential for a clear understanding of the extent to which the validity of a sampling method may be dependent on the behaviour of a particular vector species.

1. Catch of day-time resting populations indoors

Sampling of the anopheline population in their day-time resting places, particularly in habitations, has long been a standard method in the study of malaria vectors. This method has been particularly popular with some highly domestic species which feed indoors to a considerable degree and remain resting there by day, provided the house offers suitable, dark, day-time resting places. With such species a standardized day-time catch in suitable houses has not only provided all the original information about the local and geographical distribution of vector species as well as seasonal abundance and short-term fluctuations, etc., but has provided the sole sample of the population on which such important indices as infection rates, human blood ratios, and more recently age-grouping, have been based.

With such samples of the house-resting population, there are three limitations which have long been appreciated. Firstly, the limitations of the methods used in enabling the complete house-resting fraction to be detected; secondly, the limitations of the method in estimating the house-entering population, and, thirdly, the limitations of the house-resting collection as an index of the vector population as a whole. With regard to the first point, the methods used for many years were based on a careful search by means of a flashlight of all suitable resting places indoors, and subsequent capture of resting mosquitoes by means of catching tube or sucking tube. These methods were often surprisingly productive, especially in the hands of trained catchers and with reference to suitable "good" houses. When careful search in such selected houses failed to reveal resting adults by day, vector scarcity or absence was usually assumed. The introduction of the use of pyrethrum spraying indoors, in which dead or "knocked down" mosquitoes are collected on sheets previously spread over the entire floor and other horizontal surfaces, early revealed that this technique almost invariably yielded a much higher collection of mosquitoes than could be recorded by even the most thorough visual search and hand catch. This discrepancy was particularly marked at low densities in which space spraying could frequently reveal numbers of resting vectors in houses declared negative by hand catchers. The limitations of the hand catch as a sampling method are particularly marked in high-roofed native houses where so many of the resting mosquitoes settle in situations not readily accessible by hand catch, even when a light-weight ladder is used. The method now recommended for sampling the house-resting population is hand catching (in order to obtain live, healthy mosquito material and to determine the exact resting places), followed by pyrethrum space spraying in the same rooms or houses. This general recommendation also deals satisfactorily with certain flimsy, open types of habitation in which the space spraying method itself has certain limitations.

With regard to the second point, it has long been recognized that certain species of *Anopheles* bite indoors, but only remain resting indoors by day to a limited extent. With such species, the house catch, although used occasionally, has usually been abandoned as a routine in favour of other methods. For example, it has been shown that with *An. aquasalis*, a widespread vector in Central and South America, the day-time resting population indoors is only 1·3% of the total feeding there by night (Senior-White 1951, 1952). *An. aquasalis* is one of these species which exists in enormous numbers in certain areas or at certain favourable periods of the year. In such cases the 1·3% fraction remaining indoors may amount to a high house catch, giving the impression of "domesticity". At the other extreme, mosquitoes may be extremely difficult to find resting by day indoors at a time when there is still a high vector population as a whole. In the case of *An. albimanus*, another Central

American vector, it has been shown in one area that, of each 100 females entering a house at night, only about eight remain indoors on the following day; of that eight only about one may be detected by visual searching and hand catching. These figures are based on a series of three catches, viz. (a) the use of the window trap to trap the females escaping from an occupied hut at night and at dawn; (b) this is followed by hand catching in the house, which in turn is followed by (c) space spraying in the same house to detect the total resting population. (Muirhead-Thomson and Mercier, 1952.)

While these two cases are well-known extreme examples of species in which the day-time resting population is only a small fraction of the house-entering population, it is being increasingly realized that, even with the so-called "highly domestic" species normally abundant indoors by day, the house-resting population may give only a variable and incomplete idea of the total house entry. By means of the window-trap or exit-trap technique (see p. 44), it has been shown that, even in houses providing ideal, day-time resting places, a variable population may leave the house at dawn after feeding. A further variable proportion, with the blood meal half digested and the ovaries half developed, may leave at dusk of the following evening (Muirhead-Thomson, 1951; Gillies, 1954, 1955). With regard to the third point, it appears that, in view of these variations and variables noted above, even under ideal conditions with a highly domestic species and suitable day-time resting places indoors, it is unlikely that the house-resting population provides anything more than a very approximate and unreliable indication of the vector population as a whole.

2. *Day-time captures in outdoor resting places*

In view of the points made above that some vector species are poorly represented in the day catch indoors, considerable attention has been devoted to the search for natural resting places outdoors, and to the application of this knowledge to develop a routine sampling method. Routine collection of *An. minimus flavirostris* in shady stream banks and ravines in the Philippines as a recognized standard method was established over 30 years ago (Russell, 1931; Russell and Santiago, 1934). More recently a further impetus has been given to the development of regular methods of sampling outdoor resting populations by the fact that the irritant effect of DDT deposits in houses greatly reduces the attraction of such places as indoor resting sites. Many of the mosquitoes entering and feeding in these treated houses are irritated before they have absorbed a lethal dose of insecticide and leave the house to seek outdoor resting places. The methods of sampling outdoor-resting populations of anophelines may be divided conveniently into (a) the use of natural resting places, e.g. vegetation, shady ravines, crevices, caves, etc. and (*b*) the use of artificial shelters specially constructed to form attractive concentration sites.

(*a*) Natural Resting Places

Two good examples of this method come from the American Region. In Trinidad the low day-time catches of *An. aquasalis* in houses and stables, combined with the very high night-biting population, all pointed to the existence of extensive outdoor resting by day (Senior White, 1951, 1952). Resting places were finally found in low scrub, and a sampling routine was worked out on a direct search and capture basis. As many as 78 females have been recorded per man-hour by this method, the mean catch covering a wide range being 8·3 females per man-hour. In Trinidad this outdoor-resting population not only provided the basis for estimation of local incidence and seasonal abundance, but also provided ample material for the analysis of infection rates, blood-feeding preferences and age grouping. An example of a comparison between day catch in an untreated house and in the surrounding scrub was carried out in one of the villages, with the results shown in Table V.

TABLE V

Comparison of day-time catch of *Anopheles aquasalis* in untreated house in Trinidad, with simultaneous catch in the scrub surrounding the village (after Senior-White, 1952).

	June	July	Aug.	Sept.	Oct.	Nov.	Dec.
An. aquasalis females per man hr resting outside	6·2	16·8	9·2	4·9	0·7	0·6	1·2
An. aquasalis House-resting females per man hr	1·4	1·7	0·8	0·8	0·6	0·2	0·2

Studies carried out in Eastern Colombia showed that the grass of the savannah formed not only the diurnal resting place of the local anopheline mosquitoes, but also the true habitat of several species (de Zulueta, 1950, 1952). Routine sampling was carried out by means of a square muslin-covered tent, 2 m long by 2 m high, kept in position by four rods. The grass underneath was then sprayed with repellent. Tent captures of this kind were made in series along a straight line, and compared with stable-trap captures and with captures of mosquitoes flying at night (by means of two large butterfly nets attached to poles extending at each side to a vehicle travelling at about 30 km per hour).

(b) Artificial Outdoor Resting Places

In order to provide attractive concentration sites to facilitate the sampling of outdoor-resting populations a wide range of modifications have been worked out. The original "earth-lined trap" used for *An. minimus flavirostris* in the Philippines has a modern counterpart in the "box shelter". The more recent "artificial pit shelter" has proved a very attractive concentration site for several species of African and oriental *Anopheles* (Muirhead-Thomson, 1958, 1960a, 1963). In addition such methods as barrel shelters, horizontal culvert shelters and "nail keg" shelters, have all played a part in sampling the outdoor-resting populations of *Anopheles* (W.H.O. 1963). While these methods have been extremely useful with some vector anophelines, there are still species which do not appear to be attracted to the present wide range of artificial resting sites available, although large outdoor-resting populations are known to exist. *An. sundaicus* in the oriental region and *An. albimanus* in the neo-tropical region are in this category. The use of these artificial outdoor resting places for routine sampling has found its fullest expression in many malaria eradication programmes where conventional methods of sampling indoor-resting populations have revealed severe limitations after the house has been treated with insecticide. In theory, the outdoor-resting population should constitute a more representative cross-section of the population as a whole than perhaps any other method. This is because this method of sampling, if carried out well away from habitations, would not appear to be biased by the presence of man or animals (as in houses and stables) which might attract mosquitoes in differing degrees. However, despite the considerable advances which have been made in using these artificial concentration sites to estimate changes in composition of the population (with regard to age grading, proportion engorged, etc.), or to estimate density changes determined by season or the influence of total spray coverage, interpretation of these outdoor samples is still in a comparatively early and uncritical stage.

As the nature and availability of outdoor resting places in general may vary widely according to seasonal changes in climate and vegetation cover, it follows that the attraction of artificial shelters, which are in competition with natural resting places, must also be subject to certain variables according to season and topography. A striking example of a seasonal change in the relative validity of two outdoor sampling methods is provided by observations on *An. rufipes* and *An. pretoriensis* in Southern Rhodesia (Muirhead-Thomson, 1960a). Neither of these species was associated with houses or habitations, but both could be taken regularly in certain narrow, shady, ravines. Both species were also recorded in artificial pit shelters which had been constructed locally for the routine sampling of *An. gambiae* and *An. funestus*. An analysis of routine captures carried out regularly throughout all months of the year in both types of outdoor sites revealed interesting differences, as shown in Table VI.

TABLE VI

Number of blood-fed and gravid females of *Anopheles rufipes* and *An. pretoriensis* and taken in parallel collections in artificial pit shelters and in natural ravines according to season (Southern Rhodesia), (Muirhead-Thomson, 1960a).

| | | An. rufipes | | |
		Pit shelter	Natural ravine	Total
Dry season (July–Oct.) 22 collections	No. found	108	278	386
	% of total	28%	72%	
Rainy season (Feb.–March) 6 collections	No. found	3	410	413
	% of total	1%	99%	

| | | An. pretoriensis | | |
		Pit shelter	Natural ravine	Total
Dry season	No. found	51	163	214
	% of total	24%	76%	
No. found	Rainy Season	1	253	254
% of total		0·5%	99·5%	

The figures for both species show that the relative attraction of the two sites differs considerably between the dry season and the rainy season. In the dry season the artificial pit shelter recorded 28% of the total outdoor catch of *An. rufipes*, but only 1% of the total in the rainy season. With *An. pretoriensis* the percentage fell from 12% in the dry season to 0·5% in the rainy season. It is clear that samples restricted to the artificial pit shelter might be useful in the dry season for both these species, but in the rainy season might fail completely to reveal the existence of these anophelines, even at high population densities. Presumably such differences might also be expected, irrespective of season, in areas or localities which differ in the degree of natural vegetation or availability of natural resting places, in such a way that the use of artificial concentration sites might well be feasible or valid in one locality but quite misleading in another.

3. *Catches of female mosquitoes biting human and animal bait indoors and outdoors*

Catches on human and animal bait have been used extensively in studies on anophelines, very often with the dual objective not only of securing a sample of the vector population, but also with the idea of gleaning information about host preferences, blood-feeding habits, and degree of vector/man contact. In this review we are primarily concerned with the validity of these methods from the sampling point of view, although naturally the host-preference aspect must also be taken into account as it may be a factor determining the use and limitations of such sampling methods. A good example of a bait catch providing the main sampling method for routine use is shown by *An. albimanus* in Jamaica. In that area of its distribution, the vector could not readily be taken in houses, nor were the outdoor resting places known. However, the existence of a very sharp sundown peak of biting enabled a routine to be established for collecting the *An. albimanus* biting a donkey outdoors for thirty minutes at sunset. At the most favourable season of the year several hundred females could be taken biting in the course of a single catch by this method (Muirhead-Thomson and Mercier, 1952). Collections of anophelines biting bait (particularly human bait) are being increasingly used in malaria eradication programmes where vectors in areas under total spray coverage may be difficult to find by other sampling methods based on indoor-resting catch. In some places the results have been illuminating, in others disappointing. The general method has been used rather sporadically in special investigations, and it is comparatively rarely applied on a sufficiently uniform basis to be classified as a routine sampling method. In order to qualify as a recognized sampling method as distinct from an empirical index of man-biting activity, considerable uniformity would be necessary as regards (*a*) time and period of exposure; (*b*) number and type of hosts, and (*c*) location indoors or outdoors. Clearly, with regard to (*a*), the time and period of exposure must be in accordance with the known biting activities of the vector. With *An. gambiae*, for example, human-bait collections in the first quarter of the night frequently give disappointing results because the main biting activity of that species normally takes place in the latter two-thirds of the night. Perhaps for this reason the use of standard human-bait catch in the Pare Taveta malaria control scheme in East

Africa proved to be the least satisfactory of several sampling methods used in evaluation of the spraying operations.

Uniformity of host as to type and number has also been disclosed as a vital factor determining the validity of this sampling method. Early observations on mosquitoes in East Africa showed how the attraction of human bait increases enormously with increase in the number of bait-boys exposed (Haddow, 1942). Accordingly, the use of a conventional one- or two-man bait might prove inadequate or misleading in circumstances where eight–ten human bait would be necessary for sampling a vector at low densities. The adequacy of the human bait exposed is also liable to be influenced by the animal competition in the vicinity. For example, experiments were carried out on *An. melas* in West Africa in which the attraction of a fixed number of animal hosts (three to four goats, one pig and one cow) was compared with an increasing number of humans from two to four, and finally to six. Under those conditions the proportion of mosquitoes attracted to the animal hosts fell from 62% to 10%, and finally to 2%, respectively, of the total (Muirhead-Thomson, 1951).

Observations in Jamaica also indicated how dependent the validity of the human-bait catch is on the presence of deviating animals near by. The attraction of two separate family groups was compared with that of a tethered donkey and then with donkey plus cow, with the results shown in Table VII.

TABLE VII

Relative numbers of hungry female *Anopheles albimanus* attracted to human bait (family groups) and to alternative animal host near by. Observations made during 30-min peak of biting outdoors at sundown (Jamaica) (after Muirhead-Thomson and Mercier, 1952).

		Number of *An. albimanus* caught	% attracted to animal host
Expt. 1.	Human group A (2 adults and 6 children)	96	
	1 donkey (tethered 10 yd away)	86	47%
Expt. 2.	Human group B (2 adults and 4 children)	119	
	1 donkey	126	49%
Expt. 3.	Human group A (2 adults and 6 children)	127	
	1 donkey	307 ⎫	80%
	1 cow	200 ⎭	

These figures show strikingly how the roughly equal distribution of bites between human bait and the single animal is violently upset by the addition of one more animal, the proportion attracted to men falling to 20% of the total.

With regard to the use of both indoor and outdoor bait catch, some observations on anophelines have shown that, with some species, seasonal variations in the degree of indoor versus outdoor biting may necessitate special care in using such sampling methods to indicate general trends in population density. For example, a series of catches of *An. albimanus* in Jamaica were made on a donkey tethered outdoors, while simultaneous collections were made in a donkey-baited trap (hut plus window-trap catch). In the dry, summer season the findings were very uniform in showing that the amount of feeding indoors was less than 2% of that feeding outdoors. The results were consistent enough to suggest an overwhelming preference on the part of *An. albimanus* to feed outdoors. However, when the observations were continued in the cool, rainy season, there was a sharp increase in the numbers biting indoors without any corresponding increase in the numbers biting outdoors. Further observations with human family groups sitting outdoors at night, and then moving indoors and sleeping there, showed that in that season the outdoor biting was only slightly in excess of the indoor biting. From these results it appears that restriction of the sampling to indoor biting would have given misleadingly low results at a time in the dry, summer season when anophelines were abundant biting outdoors. In the same way, restriction of samples to outdoor biting might well have underestimated the great increase in vector density in the rainy season, this increase being mainly reflected in the indoor sample.

There is evidence that, where reasonable uniformity as to type and number of host, time and locality has been established, the routine bait catch can be used to indicate approximately the seasonal trends in anopheline populations. For example, the seasonal density of *An. gambiae* and *An. melas* was investigated in Liberia by means of regular all-night catches on a two-man bait team sitting outside under the overhanging eaves of a hut. Comparison of these figures with other sampling methods used in the same locality revealed a close agreement at all stages (Gelfand, 1955; Fox, 1957) (see Fig. 4).

4. *Trapping methods for the automatic capture of mosquitoes coming to bait or attempting to leave houses and habitations*

The use of man or animal-baited traps has developed along several different lines in malaria studies. For the purposes of this review it would be convenient to group them in two categories:

(*a*) The use of simple artificial traps in the form of a large box or cage occupied by human or animal bait. Mosquitoes seeking a blood meal have easy access to this trap by means of horizontal slits or louvres, but are prevented from leaving by the mosquito netting or wire gauze which forms a large part of the wall area. These traps differ in principle from those in category (*b*) below in that no special arrangements are made to direct the mosquitoes to one particular egress point when they try to escape.

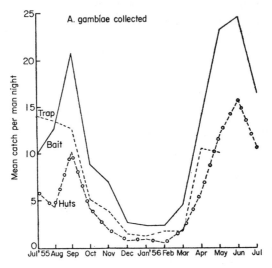

FIG. 4. Comparison of monthly catch per man-night of *Anopheles gambiae* by each of three collecting techniques (after Fox, 1957).

(*b*) The use of fixed house or hut-type traps built according to local design and materials in which mosquitoes, which attempt to leave the hut after entering or feeding, are attracted to the light coming in at one part of the house where they are trapped by netting screens (Shannon trap or verandah traps), or attracted to the light coming through a restricted window opening (usually 1 ft sq.) over which a detachable lobster-pot type of window trap is fixed. This latter design is the well-known experimental hut or trap hut, normally occupied by human or animal bait.

a. Simple Artificial Traps

The development of simple traps of this kind has taken place in two rather different directions. One type takes the form of an artificial stable trap, provided with a roof, and with walls constructed of wire mosquito gauze on a wooden framework. Mosquitoes gain access through horizontal slits or louvres in the walls, and are hand collected inside the trap on the following morning. Baited with horse, donkey or calf, traps of this

kind ("Magoon" traps, "Egyptian" traps, etc.) have been used extensively in some countries for estimating population changes and seasonal
trends in vector densities. They are considered to be satisfactory in areas
where the particular local vectors of malaria occur at very high densities,
e.g. *Anopheles albimanus* in Central America, and *An. pharoensis* in Egypt.
Their performance at low vector densities is however more debatable:
as egress is non-directional there is likely to be some escape of mosquitoes through the ingress baffles, and, in addition, the performance of the
trap will be influenced by the extent to which a particular species of
mosquito will actually enter a roofed or partly-roofed structure in order
to feed.

A rather different development of the same principle is the baited net
trap. This takes the form of a large mosquito net, the corners of which
are supported by four poles, the net being suspended in such a way as to
leave a gap of about 12 in. between the lower edge and the ground.
Either man or animal can be used to provide the lure or bait. The net
trap is easily portable, and can be set up and dismantled quickly and
easily. This type of trap, with no solid walls or roof, has given good results with those anophelines which prefer to feed outdoors in the open.
It has also the advantage of retaining its efficiency at low vector densities (see page 48). Mosquitoes trapped inside this suspended cage
can conveniently be hand-collected at dawn, but a useful variation is to
construct an artificial pit shelter inside the cage (or set up the cage over
an existing pit shelter) and thus provide an attractive aggregation site
into which mosquitoes, unable to escape from the cage, will move from
sunrise onwards (Hadjinicolau, 1963). Somewhat the same principle of
sampling has been used in Malayan studies in a very different context,
namely, investigations on the anopheline vectors of monkey malaria
(Wharton *et al.* 1963). It has long been recognized that there are peculiar difficulties in the way of sampling mosquitoes attracted to monkeys
in their natural environment. The method eventually found to be the
most reliable was to use a modification of the original Malayan bed-net
trap (in which the human bait was protected from mosquito bites by a
bed-net) (see Chapter 4) to accommodate an expanded metal cage in
which the monkey bait was fully exposed to the bites of mosquitoes entering the trap. The trap differed slightly in design and in operation
from the net trap discussed above in that mosquitoes entered the trap
through rather wider side opening — 22 in wide in the larger of the
two models — and are manually captured inside by human collectors at
hourly or 2-hourly intervals throughout the night.

b. Experimental Huts or Trap Huts

The use of the experimental hut technique has received increasing attention in recent years in keeping with the growing need for more accurate information concerning the movements of mosquitoes in and out of occupied houses. The sampling of these different fractions of the mosquito population, combined with recording their mortality, forms a vital phase in evaluating the effect of treating houses with insecticide for the control of malaria vectors. It is particularly illuminating to see what modifications the basic design has undergone according to the different conditions encountered, or according to changing needs. In its original and perhaps simplest form, the experimental hut was a small rectangular hut with mud walls and thatch roof; it was constructed according to local design, but in such a way that the only obvious light entering the hut was through a 1-ft sq. window opening, over which a simple lobster-pot type of exit trap was fixed. The hut is occupied during the hours of darkness by one or two men or boys who provide the human bait. Hungry mosquitoes can enter the hut through the innumerable small gaps and chinks formed where the thatch eaves rests on the mud walls. From inside the hut, however, the only visible light coming in from outside, by day or by night, is through the window opening. Mosquitoes leaving the hut, usually after feeding on the occupants, are attracted to this opening, which is still clearly visible on the darkest night, and are automatically trapped in the window cage, which can then be detached or replaced.

In this kind of hut, samples are obtained of the fraction of mosquitoes which remain resting indoors during the day, and the fraction which leaves the hut either at dusk or at dawn. Window cages are detachable and interchangeable, and can be used to check the egress of mosquitoes from the hut throughout the 24 hours. One of the earliest modifications introduced when the basic design was applied to other house-frequenting mosquitoes in different countries was to construct entry slits or baffles in the walls in order to facilitate entry of mosquitoes (Wharton, 1951; Davidson, 1953; Coz et al. 1965) (Shalaby, 1963) see W.H.O. 1963). In other cases the number of window openings with exit traps has been increased, sometimes up to 4, i.e. with one window in each wall (Hadjinicolau, 1963).

In certain favourable conditions when isolated trap huts have provided an unusually attractive focus for high mosquito populations, reversal of the normal exit traps has enabled the mosquito population to be sampled as it enters the occupied hut (Bertram and McGregor, 1956). With such ingress traps it has been possible to show that the main mosquito invasion may come from different directions according to wind direction and other conditions.

The form of the exit trap itself has undergone modification at the hands of different workers, with the slit-type of exit being used by some, and the cone-type by others. One worker, very conscious of the possibility that even the most efficient design of exit trap might still provide some obstacle to the natural egress of mosquitoes, used a simple open egress cage without any funnel or baffle. At regular short intervals a sliding door closes off the window cage allowing the sample of mosquitoes to be removed, and the cage to be quickly replaced by another. It was considered that this frequent change of exit cage, and removal of the contained mosquitoes, would ensure that mosquitoes which entered the cage unimpeded would be collected before they showed any tendency to move unobstructed back into the hut again (Rachou *et al.* 1965).

In some instances it has been found that human bait is not sufficiently attractive to the local anopheline vector of malaria to provide adequate numbers for observation in experimental huts. In Jamaica for example a simple trap hut of exactly the same design as had previously been used successfully in Africa, was baited with a donkey and proved extremely attractive to *Anopheles albimanus* (Muirhead-Thomson and Mercier, 1952). In Indonesia it was found that the human-baited trap hut or experimental hut was suitable for one of the local vectors of malaria, *Anopheles sundaicus*, but not for the other vector, *An. aconitus*. For the latter species the use of an animal bait, particularly water buffalo, proved highly successful (Soerono *et al.* 1965). Similarly in India, it was found that a bait calf attracted many more *Anopheles culicifacies* than a human bait. In practice it was found useful to adopt a design of trap hut with a low partition across the inside, the human bait/collector sleeping on one side and the calf bait on the other. In this latter example it is worth noting that this pattern did in fact conform closely to the local housing conditions where there was a prevalence of "mixed" dwellings in which man and domestic animals shared the same room and roof (Shalaby, 1963 see W.H.O. 1963).

Entomologists working with trap huts or experimental huts have long been aware of possible sources of error when using this technique to obtain an accurate sample of the egressing fraction of the mosquito population. They have realized that modifications used to facilitate entry of mosquitoes, e.g. entry slits, louvres or baffles, might at the same time provide alternative means of egress. Checks on this possibility carried out on the prototype hut described above — by lowering a very large mosquito net over the whole trap hut before dawn — did confirm that with the particular species concerned egress by means other than the window trap was negligible. However, the extension of this principle of sampling to other house-frequenting mosquitoes, and the introduction of various modifications in the way of ingress baffles and design of exit traps, demanded a constant reappraisal of this possible source of error. This was found to be particularly necessary when the trap hut design is used to evaluate the effect of house-treatment with new types of synthetic insecticide, whose effect on mosquitoes might differ in many ways from that of DDT, dieldrin and other well-known chlorinated hydrocarbons.

The whole question of hut-trap efficiency has lately been critically re-examined in East Africa and has led to the development of a verandah trap hut (Smith, 1965a, b). This consists basically of the standard type of experimental hut on to which verandahs are constructed on all four sides. By the addition of mosquito wire screens to one of the verandahs, it is converted into a trap for capturing those mosquitoes which escape from the hut by the eaves at or before dawn (as distinct from the window trap) on that particular side. With the window trap in position, the verandah trap would record the previously unassessed fraction of mosquitoes leaving through the eaves. The results showed that with *Anopheles gambiae*, of all gonotrophic stages leaving the hut, 85% of the egress occurred in the window trap, and 15% by the eaves. In the case of the prevalent house-haunting culicine mosquito in that area (*Mansonia uniformis*) there was a considerably greater egress by the eaves; of all freshly-engorged females leaving the hut, only 34% of the egress took place via the window trap.

The exact proportions will doubtless tend to vary a great deal according to the exact design and construction of the hut, eaves, ingress baffles and exit trap, but undoubtedly the verandah trap marks a new and critical approach to this aspect of mosquito population sampling.

The general type of trap hut described so far is usually designed to simulate local conditions of housing and housing material. Nevertheless, the technique does involve some artificialities in construction, and in having the number and movement of bait animals strictly controlled. There still remains some uncertainty therefore as to how far the sampling data based on such an experimental hut are really representative of conditions in normal, occupied, rural habitations, where there are so many uncontrollable variables. Sometimes, the technique can be applied directly to certain well-constructed occupied houses which already possess a window opening over which a cone-type exit trap can be fitted. In some areas, however, the local houses are of closed contruction, with solid walls and no window openings. In such cases the modification adopted has been to replace existing doors with well made close-fitting ones, into which a 1-ft sq. opening has been cut to form a window opening for the exit trap. When the trap is not actually being used, the square piece of wood, hinged to the top of the opening, can be closed to form a shutter and can be bolted from inside (Service, 1963). In such tightly closed houses the window trap can be inverted to function as an ingress trap as well. In all cases involving sampling anopheline mosquitoes in normal occupied rural habitations, success or failure may be entirely dependent on the good-will and co-operation of the occupiers.

Somewhat similar developments are taking place in Central America where a very prevalent type of rural habitation has walls constructed of a lattice of sticks providing innumerable openings through which the light penetrates. Under such conditions the conventional exit trap is of limited use, as mosquitoes can leave or enter by any of the many alternative routes. As the anophelines in some of those countries tend to leave habitations at dawn to a great extent, the method developed to sample the house-frequenting population has been to construct a whole demountable battery of exit cages which can be set up to cover the entire wall of a house if necessary.

5. *Combination of different sampling methods*

In view of the now recognized limitations of any single sampling method used alone, there is an increasing tendency in anopheline studies to use a combination of two or more sampling methods. It is now normal practice to supplement indoor sampling of the day-resting population by outdoor sampling of resting adults and by night-time catches on human or animal bait. Although this development has arisen under the pressure of malaria eradication projects in which more accurate methods of entomological evaluation are urgently required, it is a development of considerable significance in the wide context of sampling procedures in general. In some cases two sampling methods, employing different principles, have been used simultaneously over a sufficiently long period to form a reasonable idea of their comparative validity. In a formerly BHC-sprayed area in Southern Rhodesia, even after spraying had been withdrawn for two or three years, the main vectors, *An. gambiae* and *An. funestus*, were found to be feeding and resting almost entirely outdoors with very little evidence of the high degree of house contact which existed prior to spraying (Muirhead-Thomson, 1963). One entomologist worked out a sampling method based on the use of artificial outdoor pit shelters constructed on a standard pattern. A later entomologist concentrated on working out a trapping technique with human and animal bait. Finally, both methods were carried out simultaneously throughout a whole season in the same village (see Fig. 5) (Hadjinicolau, 1963).

The results indicate that in the period October to March, which corresponds to the cooler rainy season of the year, much higher numbers of both *An. gambiae* and *An. funestus* were recovered from the baited net trap than the pit shelters. This is particularly striking in the month of maximum mosquito density — March — at the end of the rainy season. In the dry hotter season, from April to September, the pit shelters become as productive — on occasions more productive — than the baited net trap. This comparison also reveals that, while the pit shelter is a useful sampling technique for some vectors, it is obviously unsuited for other vectors such as *An. coustani* (and *An. squamosus*) which are always more readily taken by bait than by pit shelter.

The present trend towards a wide spectrum sampling of anopheline mosquito populations is well illustrated by the two following examples.

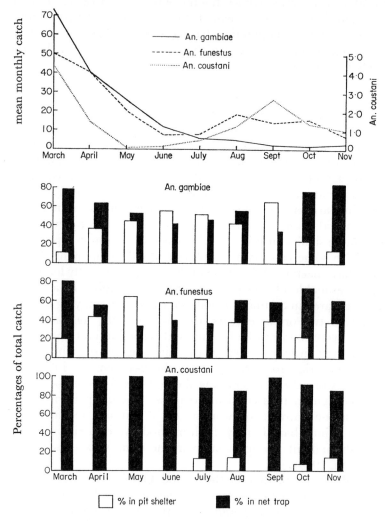

F IG . 5. Seasonal abundance of three species of *Anopheles* viz. *An. gambiae, An. funestus* and *An. coustani* in Southern Rhodesia (as indicated by mean monthly numbers per collection in eighteen pit shelters and one baited net trap), together with partition of mosquitoes between net trap and pit shelter according to season. (Adapted from Hadjinicolau, 1963.)

In entomological evaluation of malaria eradication operations in Mozambique, Africa, the methods used for sampling the two main vectors of malaria, *An. gambiae* and *An. funestus* embodied the combined experience of many different entomologists who have studied these species, so closely associated with man and his habitations (W.H.O. 1963). Sampling methods used were as follows:–

i. Vector populations resting indoors by day as judged by the pyrethrum space-spray collection.
ii. Outdoor day-time resting populations by means of artificial pit shelters.
iii. Baited net catch in which the trap net can be baited with man indoors or outdoors, and with cow outdoors.
iv. Window trap catch to sample mosquitoes leaving the house after feeding or resting.

In methods i, ii and iv a record was also kept of the female vectors captured with regard to "unfed", "fed" and "gravid".

The second example is provided by investigations carried out in Malaya under very different conditions (Moorhouse and Wharton, 1965). These studies were concerned with three known malaria vectors in Malaya, and three suspect vectors. Four of these six species of *Anopheles* are forest or swamp-forest mosquitoes which show widely different degrees of contact with human settlements. Of these, two species are rarely taken outside the forest fringe, while the other two readily fly into human settlements from the forest at night. Four sampling sites were selected representing a range of environmental conditions, such as swamp-forest edge, coastal kampong (human settlement), and foothill area. At each site the following methods were used:–

i. Man-biting catches inside houses.
ii. Man-biting catch outside house.
iii. Man-baited net trap emptied once every hour.
iv. Man-baited net trap emptied every 2 hours.
v. Pyrethrum knock-down catch inside house.

In addition, a continuous 24-hour catch on human bait was carried out at both the swamp-forest sites.

The value of this broad front approach to sampling was very evident. The 24-hour bait catch for example not only revealed large numbers of a particular species, *An. roperi*, hitherto regarded as uncommon in Malaya, biting man in the forest shade, but also revealed a remarkable peak of biting activity in the hour before sunset, in which period over 50% of the total biting took place. These workers considered that although the 24-hour catch is rather too laborious for routine use, it has a fundamental role to play in the design of sampling methods. The range of sampling techniques also revealed that of the several species which may enter houses to feed, only *An. campestris* regularly rests indoors by day. The net trap, normally occupied by human bait protected by mosquito netting, was found to sample the mosquito population entering the settlement at night, while the actual bait or biting catch sampled the feeding fraction of the population. The use of both methods revealed that a considerable proportion of the anopheline population arrived in the settlements early in the evening, but rested for a significant period before feeding.

In investigations on forest-dwelling mosquitoes in south and central America, and in Africa, extensive use has been made of tree platforms to study the biting activity of mosquitoes on exposed human bait at different levels from ground to forest canopy. The bulk of that work — as will be described in the next chapter — has been concerned with culicine mosquitoes, but in the course of that work a great deal of new information about anopheline mosquitoes has also come to light, some of which concerns major vectors of malaria such as *Anopheles gambiae* in Africa and *An. darlingi* in central and south America.

Out of the great wealth of new information, it will be sufficient for the moment to quote a particularly illuminating example — somewhat similar to that of *An. roperi* noted above. In studies in the forests near Belem in Brazil, captures were made simultaneously on human bait at ground level, and on platforms 5, 10 and 15 metres above the ground (Deane *et al.* 1953).

In the case of the main malaria vector *An. darlingi*, 50% of the mosquitoes were taken at ground level compared with 12% at 15 metres. With two other species of *Anopheles* however, *An. mediopunctatus* and *An. shannoni*, 60% of the biting was recorded at the highest platform compared with only 3–4% at ground level. Previous routine entomological surveys carried out in houses and on exposed bait at ground level had revealed comparatively few adults of these two species, even though their larvae could be found readily in ground pools and rivers which provided the breeding places.

CHAPTER 5

CULICINE MOSQUITOES

As has been pointed out earlier in this report, methods of sampling mosquitoes have tended to develop along different lines according to the different diseases, or groups of diseases, they transmit. This dichotomy is due, not so much to fundamentally different principles involved, as to the fact that specialists in one particular insect-borne disease inevitably tend to become isolated from parallel developments in the study of other diseases. This applies even within the more restricted field of culicine vectors of disease where independent methods have tended to develop according to the different ideas of isolated groups of workers.

SAMPLING OF MOSQUITOES IN SYLVAN YELLOW FEVER STUDIES

Undoubtedly the most extensive sampling of culicine mosquitoes both with regard to absolute numbers and to wealth of species is that which has been carried out in Central and South America and in Africa, originally in connection with studies on jungle or sylvan yellow fever, but latterly more concerned with studies on the increasing number of arbor viruses being recorded. The ultimate choice of sampling method in these studies has been largely influenced by the original objective of this work, namely, the extensive screening of sylvan mosquitoes for virus, and the investigation of the vectorial link between sporadic infection in the human host and the reservoir of infection in arboreal monkeys or other forest animals. For the understanding of this latter aspect it was necessary to find out much more about the biting activity of the various culicines at different times of the day and the night, and at different levels in the forest according to the habitats of the animals forming the reservoir of virus.

For this and other reasons the bait catch has been the method of choice, and has been developed to a high pitch of refinement, not only with regard to complete representation in time — the 24-hour catch — but also with regard to complete representation concerning height from the forest floor to the forest canopy (Haddow, 1954). In Central Africa the groundwork for the standard method finally adopted had been laid in previous critical studies on the attraction of *Anopheles* to human bait, in which it was shown that the degree of attraction varied considerably from individual to individual, and was also influenced by the number of individuals forming the bait and by whether the bait was washed or unwashed (Haddow, 1942). These and other human variables dictated a basic bait of not less than three boys as a convenient number to counteract individual variation in attraction, and at the same time to provide a considerably greater source of attraction than a single human. The bulk of the catch was carried out with units of three boys, one boy at a time actively catching off himself and the other two. With a complete unit of five, a staggered unit system could be used, each boy acting as catcher for the first hour after coming on duty. When he had finished catching he had two hours' light duty, then two hours' rest. Broadly speaking two main classes of biting activity were defined. In the first group there is, at some time during the 24 hours, a single short, very pronounced wave. This may be so sharp as to be mainly restricted to the single hour after sunset, for example *Aedes africanus*. A sub-type of this group is where two distinct waves occur. In the second group the phase of biting is prolonged, either nocturnal or diurnal, and the biting cycle is usually irregular.

Out of the great wealth of material emerging from these studies, we must restrict ourselves to those salient features which bear on the sampling aspect. Although this work was not primarily designed to sample or estimate the population as a whole or to follow seasonal fluctuations in numbers, much of the incidental information on this point is very instructive.

With regard to seasonal changes it was found that in general the pattern of the 24-hour biting cycle remained much the same, but at lower numerical values in the dry season. However, with *Anopheles gambiae* which happened to be taken in large numbers in some of the forest areas, catches restricted to the early part of the night (a period of low activity) appear to give unusually low figures in the dry season because the proportion feeding in that part of the night is only about 2·2% in the dry season as compared with 6·7% of the night catch in the rains. As the author says "Small seasonal changes in the form of the biting cycle may affect quite profoundly the estimate of seasonal incidence unless the whole 24-hour period is used" (Haddow, 1954). These studies also showed that loss of biting cycle may occur in the following circumstances:–

(a) the use of bait not normally attacked by the mosquitoes concerned; several species only bite man or primates casually;
(b) catching in an unfamiliar environment;
(c) cases where mosquitoes are travelling, for example between breeding grounds and some distant biting centre such as a village (Haddow, 1961).

Another interesting example, from outside the field of culicine studies, illustrating the influence of extraneous conditions on biting activity and on the validity of sampling based on the use of the bait catch, is provided by studies on *Chrysops*, the tabanid vector of human loiasis. Sampling of *Chrysops* — a diurnal biter — is based on standard human bait exposed during two consecutive periods of the day, 8 a.m. to 1 p.m. and 1 p.m. to 6 p.m. (Duke, 1959a, b). The number of flies taken per ten boy-days during the month is used to denote the "monthly biting density". In collections at ground level, in the rain forest of the Cameroons, it was observed that the number of man-biting *Chrysops* caught by a single fly boy was increased when he sat near a wood fire. The effect of this wood smoke on stimulating biting was twice as great in *C. silacea* as on *C. dimidiata*. The biting reactions of these two species were also influenced by whether the bait was moving or stationary, *C. dimidiata* being more efficient at finding a stationary host.

Parallel studies on forest mosquitoes in connection with sylvan yellow fever in Central America (Galindo *et al.* 1950) were also based mainly on catches taken on human bait in the open, at different times and at different levels. As the jungle mosquitoes involved in the virus transmission cycle were mainly diurnal, the complete 24-hour biting cycle was not used as a routine, most catches being carried out at regular intervals between 8.30 a.m. and 5.30 p.m. one day each week throughout the year. Of particular interest in connection with the subject of the present report is the fact that these standard bait catches were used as a basis for estimating the annual cycle of abundance, the mosquito captures being recorded as "mosquitoes per ten man-hours per month". In this way the maximum monthly rates of *Haemagogus* sp (the main vectors) found in the study area in Panama were compared with those obtained in endemic centres of yellow fever in South America. In this connection the authors made a highly significant comment. "It must be noted, however, that what we record here are not truly 'densities' of the populations of these mosquitoes, but only the relative numbers attacking man. It may be that these mosquitoes may have preferred hosts for blood meals about which we have no information." Again, commenting on the low captures recorded in the dry season, they stated: "We cannot exclude the possibility that they may have other preferred hosts during the dry season, about which we know nothing".

In the Central African studies some observations in addition to the more regular routine of platform catches provided some interesting data on the validity of the bait catch under different conditions (Lumsden, 1951). Simultaneous catches were carried out throughout the 24 hours on human bait in four different environments frequented by man in settlements near the forest: (*a*) the forest floor, (*b*) banana plantations, (*c*) open space, and (*d*) hut. Each catch was divided into two components:

    i. the catch during the period that the particular environment was liable to be occupied by humans, and

    ii. when it was not normally occupied by humans, e.g. hut by day or forest floor and plantations between the hours of 19.00 and 05.00.

Preference was given to collections made during periods of normal occupancy. The results showed that some species have a predominant preference for one site over the other, e.g. *Aedes simpsoni* in plantations occupied by man, *Aedes circumluteolus* mainly in the forest-floor catch. *Anopheles gambiae* was taken in all environments, mostly irrespective of time of occupation by man.

    A somewhat similar extension of the 24-hour catch in order to represent a wider range of biting habitats has characterized recent work on *Aedes aegypti*, and other allied species of culicines in West African villages (Boorman, 1960a, b). Those 24-hour catches were made continuously on a shift system over a period of four days at regular intervals — as far as possible during the last week of each month — and were carried out in the following different types of catching station:–

    (*a*) platform in tree at 65 ft
    (*b*) at foot of tree
    (*c*) by the side of a path 400 yd from the centre of the village
    (*d*) near the centre of the village
    (*e*) inside a partly completed hut.

It is noteworthy that large numbers of male culicines were also taken at these bait catches, and that the catches were used to give an estimate of seasonal density of both males and females. The total number of each species taken each month was expressed as a percentage of the total catch of that series. The arrival of the non-biting male culicines at bait was recorded in terms of an "activity cycle" of males after the manner of the biting cycle of females.

    The regular occurrence of male culicines coming to bait was also noted in studies on the 24-hour activity cycle of domestic *Aedes aegypti* in Tanganyika (Lumsden, 1957). In that study catches were made inside a hut and on the verandah outside. The results indicated that both inside the hut and outside about two-thirds of the total biting activity took place by day. Similar observations in the coastal region of Kenya (Teesdale, 1955) revealed consistent early morning and late afternoon periodicity of activity in *Aedes aegypti*. In that work it was also observed that at the very start of a catch, when the host suddenly became available, there was usually sharp biting activity irrespective of the time the catch started. This was particularly noticeable in houses where the sudden entry of catchers into a confined space produced a sharp biting wave. In this connection it is interesting to recall that, in the classic studies on *Aedes aegypti* as a vector of urban yellow fever in the Americas, sampling or collection of adults was almost entirely confined to the house-resting population (Soper *et al.* 1943), and that it is only in recent years that sampling by bait catch has been applied to this highly domestic culicine.

In the last two years the routine platform catches using human bait have been supplemented in the Central African studies by light-trap catches at different levels (Haddow, *et al.* 1961; Corbet, 1961). In these catches males formed nearly 85% of the total culicines. Studies on *Mansonia* — which formed the bulk of the culicine catch — showed that the great majority of the females were unfed and did not contain visible eggs. No gravid females were caught. A most interesting feature which emerged was that flying females were prevalent throughout most of the night, including those species with a sharp, clearly-defined biting peak. The biting peak therefore does not appear to be necessarily determined by a corresponding peak of general activity. (See also Chapter 4.)

A fairly general conclusion from all those experiences with light traps is that they attract the phototactic part of the mosquito night-flying population, but only at times when they are not preoccupied with other activities, such as swarming, biting, and ovipositing (Corbet, 1961; Standfast, 1965).

In view of this, and of the many variable factors influencing the bait catch, increasing attention has been given to the development of non-attractant methods, such as the suction trap (Lumsden, 1958; Minter, 1961). The possible value of this sampling technique has now been thoroughly explored in the Central African studies, suction trapping being carried out at all seven levels (0–120 ft) of the specially constructed high steel tower on which so much of this work has been based in recent years (Goma, 1965). The most important conclusions from this study were as follows. Firstly, there was a surprising paucity of males of the culicine mosquitoes in these traps, suggesting that apart from activities associated with swarming, or with attraction to light, males are very inactive. There was also a paucity of gravid and engorged females, again suggesting comparatively low flight activity at those phases. Also noteworthy was the finding that very few mosquitoes were caught in these suction traps above the forest canopy as compared with the considerable number in light traps and man-baited traps.

One particularly significant point which emerges from those intensive studies in Central Africa is that they deal almost exclusively with method of trapping or sampling the adult mosquito population at periods when they are normally active or in flight. Sampling of the outdoor-resting population during periods of inactivity — usually by day — has received much less attention, and in fact surprisingly little is known about the normal sheltering sites of even such well-studied mosquitoes as *Aedes africanus*, a key species in the complex epidemiology of sylvan yellow fever. This gap in knowledge refers in particular to the regularly recurring periods between feeds, when the female's blood meal is digesting and the ovaries are developing. From an appraisal of conditions elsewhere in Africa (Service and Boorman, 1965) it is evident that there is in general a remarkable lack of information about this important phase in the lives of adult culicine mosquitoes, and that a great deal of basic research will have to be done before there is any hope of developing sampling methods at all comparable to those which have been devised for various anopheline mosquito vectors of malaria in outdoor resting places.

CULICINE SAMPLING IN OTHER VIRUS STUDIES

Observations and trials on culicine sampling in Singapore have provided some very illuminating data on the general rationale or validity of bait catches (Colless, 1959). For many years entomologists in Malaya, faced with the difficulty of finding adequate anopheline malaria vectors in houses, have been accustomed to use the "Malayan human bait trap" for routine catching. This is a large mosquito net, 6 × 6 × 6 ft, with doors that are closed by flaps which can be dropped by a simple device operated from inside. Inside this large net is another smaller net under which the operator, who acts as bait, sleeps. At intervals he rises, lowers the outside flaps, and catches the trapped mosquitoes.

Realizing that one of the obvious defects of this design for culicine sampling was that some species, unable to bite, would quickly fly out, a modified trap was designed. This consisted of a large net only with a single open doorway, and with two or three men collecting all night. Among the first rather unexpected results was the finding, by means of the precipitin test, that some of the blood-fed mosquitoes taken in the trap had evidently fed elsewhere before entering the trap. In addition, large catches of all stages were regularly taken in an unbaited trap of this type. The results of catches revealed two different types of reaction to these baits and traps. *Culex tritaeniorhynchus*, for example, enters the traps in numbers, but only a small proportion — less than 1% — actually fed on the human bait. *Culex fatigans* on the other hand was not strongly attracted to the human bait, but once it found its way into the cage it fed readily on the bait. From the point of view of host preference studies it appeared that figures based only on the numbers attracted to the baited trap might be misleading, and that in fact unbaited traps proved much more useful for obtaining relatively unbiased samples of engorged mosquitoes for precipitin testing. It is clear that in the wider context of population sampling by means of bait catch these findings have considerable significance.

Other workers in Malaya (Reid, 1961; Wharton, 1951) have provided an interesting comparison of the relative proportion of different mosquito species, both culicine and anopheline, attracted to the same bait under two different trapping conditions, viz. the Malayan type of net trap and the window trap hut. When different species are offered a choice of host (two men versus one cow) in two net traps, an "attraction ratio" can be established for each species indicating the partition between the two types of host (see Table VIII). When the mosquitoes

TABLE VIII

Relative attraction of Malayan mosquitoes to human and animal bait according to capture method adopted. (Comparison of the man : calf attraction ratios obtained with net traps and with window trap huts) (after Reid, 1961 — Wharton, 1951).

Species of mosquito	Net traps			Window trap huts		
	2 men	1 calf	Ratio men : calf	1 or 2 men	1 calf	Ratio men : calf
Culex pipiens fatigans	909	531	1·7 : 1	442	111	4 : 1
Culex "annulus"	166	353	1 : 2·1	48	942	1 : 20
Mansonia uniformis	243	687	1 : 2·8	16	582	1 : 36
Anopheles "hyrcanus"	6	36	1 : 6·0	0	61	1 : > 61
Culex gelidus	688	5 328	1 : 7·8	25	2 658	1 : 106
Anopheles vagus	12	985	1 : 82	0	122	1 : > 122

are offered a choice of the same man/animal hosts in window trap huts, the bias towards the preferred host — man in the case of *C.p. fatigans* and calf in the case of the other five species — is greatly intensified, up to ten-fold or more with some species. It appears, therefore, that even with a standard choice of host designed to attract a range of species with widely different host preferences, the validity of the bait catch as a sampling method can still be greatly influenced by the actual design or technique of trapping. A parallel critical appraisal of all the variable factors affecting the bait catch and its interpretation has been made by a group of workers in quite a different part of the world, namely in Nigeria, West Africa (Service and Boorman, 1965). Experience there, with both culicine and anopheline mosquitoes, has shown clearly that in addition to such factors as locality, time of day and year, nature of bait and so on, the question of variation between individual traps must be taken into account when interpreting bait catch data, quite apart from differences in the actual type or design of trap. Sampling may be biased by trap idiosyncrasies which are highly significant for the mosquito although they may not be readily apparent to the human eye.

Factors affecting the use of baited traps were also studied critically in the ecological studies on Japanese B. encephalitis virus in Japan (Scherer *et al.* 1959; Buescher *et al.* 1959). In those studies considerable use was made of two kinds of trap; small ground traps baited with birds or chickens and larger Magoon-type traps or modifications which could be baited with man or with larger animals. In these comparative trials the number of animals was adjusted so as to give approximately similar body areas, for example, more small birds were used per trap than large ones. The experiments took the following form:–

(a) comparison of seven different bird genera;
(b) comparison of Black Coated Night Heron (the most prevalent local bird), pigs and man as attractants;
(c) effect of altitude on mosquito recovery, with baited traps up to 50 ft above ground level;
(d) effect of microhabitat and other trap variables.

Among many other points of interest these observations showed that the Heron-baited trap which was very satisfactory at high vector densities revealed certain limitations at low end-of-season densities, at which point the pig-baited trap became a more sensitive indicator of *Culex tritaeniorhynchus* populations. It was also observed that mosquitoes still entered traps empty of bait — as in the Singapore experiments referred to above — but the numbers declined after three days. The analysis of catches also revealed the important point that the engorgement rate of mosquitoes attracted to the bait can be quite independent of the attraction rate to that particular bait. (See also Colless, 1959a, b; Nelson and Chamberlain, 1955; Dow *et al.* 1957.)

In striking contrast to these well-established sampling conclusions worked out in Japan, it was found that in somewhat similar encephalitis studies in Taiwan the reactions of *Culex tritaeniorhynchus* to different sampling techniques followed a rather different pattern (Hu and Grayson, 1962). In the latter studies three different methods of sampling were tried out:–

(a) cattle bait outdoors;
(b) modified Magoon trap (as used in the Japanese studies);
(c) light trap.

These observations showed that the modified Magoon trap, so valuable in attracting *Culex tritaeniorhynchus* in Japan, gave poor results in Taiwan. This species was reluctant to enter these traps, whether baited with pig, man or ox, at a time when it could be taken abundantly by hand catch on ox bait outdoors (Table IX).

TABLE IX

Mosquitoes caught in a Magoon trap operated 24 h/day, compared to individual collection from oxen for 3 h after sunset at Chung-ho village, Taiwan (after Hu and Grayson, 1962).

		Numbers of Mosquitoes					
		Magoon Trap			Hand-caught from oxen		
	Bait	*Culex tri.*	*Culex pipiens fatigans*	*Anopheles sinensis*	*Culex tri.*	*Culex pipiens fatigans*	*Anopheles sinensis*
18–30 June	Pig	25	8	20	6 796	73	2 722
8–11 July	Ox	121	71	2	3 434	0	975

These studies also provided additional information directly bearing on the general problem of using bait for sampling, namely that the number of mosquitoes attracted to an ox or a water buffalo did not vary significantly from one animal to another.

In standard biting catches on human bait, especially those catches which are extended throughout the 24 hours, it would seem reasonable to assume that all vector species, whose importance depends on their biting contact with man, would be readily taken. In investigations on arthropod-borne virus in Togoland in South Africa it was found that one of the most important species, as judged by virus isolations from natural samples and by laboratory transmission experiments, was *Aedes* (*Banksinella*) *circumluteolus* (de Meillon *et al.* 1957; Brook Worth *et al.* 1961). Human bait catches at different times only yielded small numbers of this species. Other methods such as baited traps and light traps also proved disappointing, until the method finally adopted was to search for day-time resting *Aedes* on vegetation and grasses with a team of trained African malaria assistants. The members of the team systematically combed through a tract of bush. Mosquitoes which were disturbed were watched until they settled and then captured in glass tubes. The investigation, which was concerned primarily with virus isolation, was not continued long enough to enable one to tell how far this method could have been developed as a routine sampling method, and to what extent the catch represented stages other than hungry females or freshly engorged ones. The important point is that this method of sampling, which has much in common with that developed specifically for *Anopheles aquasalis* in Trinidad (see page 36), was successful where the more conventional 24-hour catch revealed unexpected limitations.

Brief reference has already been made (page 55) to the conventional sampling methods long used for *Aedes aegypti* in its role of vector of urban yellow fever in the Americas. For many years sampling of this classical "yellow fever mosquito" was limited to determining easily-obtainable indices of infestation, either of adults resting in houses or of larvae breeding in domestic and peri-domestic water collections. The need to have a critical reappraisal of these indices has been stimulated by recent intensification of the *Aedes aegypti* eradication programme in the United States (Tinker, 1967). So far, this reappraisal has been limited to various methods adopted for measuring infestation of larvae in breeding receptacles, such as the "average number of infested receptacles per inspected premises", or, "the percentage of inspected premises that are infested". The emphasis on larval rather than adult sampling is determined by practical considerations in large-scale survey work, and by the fact that it is considered that for *Aedes aegypti* the larval stage is the one in which the population is most readily available, and where it is most concentrated. There seems little doubt that this new approach to an old subject will lead to the development of more varied and sensitive methods for sampling the adult population as well, and that long overdue progress in this direction will be further stimulated by the increasing importance of this mosquito, and its close allies such as *Aedes albopictus*, as vectors of various new manifestations of dengue (W.H.O. 1967a).

SAMPLING THE CULICINE VECTORS OF FILARIASIS

The methods of collecting and sampling the culicine vectors of filariasis have developed along rather different lines according to whether the main mosquito studied was *Culex pipiens fatigans*, the classical vector of urban filariasis, *Aedes* and other culicine vectors of Pacific filariasis, or *Mansonia* and its allies as vectors of rural filariasis. In general, sampling of *Culex p. fatigans* populations has followed conventional methods of day-time hand-catching in houses or stables on the same lines as those long practised in the study of domestic anopheline mosquitoes (Giglioli, 1948; Rachou *et al.* 1958). By standardizing the number of houses, stables, or man-made shelters searched by day, records have been produced of the monthly and seasonal density of *Culex* populations. Considerable variations have been noted in annual densities estimated in this way, but the sampling method appears to have been the obvious selection due to the extremely high numbers of *Culex* liable to be found resting indoors, sometimes hundreds or even thousands per house.

In Indian studies on populations of *Culex p. fatigans* as a vector of urban filariasis an arbitrary collecting procedure has been widely used, namely the routine day-time collection of mosquitoes from human dwellings and cattle sheds from 7 a.m. to 10 a.m. (Wattal and Kalra, 1960). A similar routine has also been used in the studies of both *Culex* and *Mansonioides* populations in rural filariasis studies (Joseph *et al.* 1962). In these and several other studies on *Culex p. fatigans* in India and Ceylon (Chow and Thevasagayan, 1957) the high numbers of *Culex p. fatigans* taken indoors have been marked in contrast to the very low numbers yielded from a variety of outdoor resting places and natural shelters. Some interesting figures to demonstrate this point have been produced in studies on culicines as vectors of filariasis in South India (Pal *et al.* 1960). By means of routine searches indoors and outdoors, the comparative densities could be assessed on a man-hour basis. With *Culex p. fatigans* the per man-hour density indoors varied in different months from 17 to 59, contrasted with an outdoor density of 0·082 per man-hour — that is less than one adult mosquito per ten man-hours of searching. With *Mansonioides*, the vectors of *Brugia malayi*, the situation was very different. With *M. annulifera* the per man-hour resting densities were, indoors 4·23, outdoors 8·0. With *M. uniformis* the figures were, indoors 0·98, outdoors 6·62. In this series of investigations and in one or two others these day-time resting densities have been supplemented at intervals by night collections in fixed capture stations at two hourly intervals from 6 p.m. to 6 a.m.

The ease with which adult *Culex pipiens fatigans* can normally be found resting indoors, often at high densities, combined with the apparent difficulty in finding adults in outdoor resting places has led many observers to believe that no really significant fraction of the adult population exists in sites away from human habitations and associated animal shelters. However, this concept now needs drastic revision in view of the observations made by the World Health Organization Filariasis Research Team in Rangoon, Burma, in the last few years (de Meillon *et al.* 1967). They have shown that in addition to the usual house-resting adult population, a highly significant proportion of the resting population evidently uses a variety of natural shelters outdoors by day, and that this outdoor population includes females at various stages of blood-meal digestion and ovarian development.

This team has also developed additional capture or sampling techniques which are of particular interest in the present review in that they are designed to study the gravid female fraction of the population (de Meillon *et al.* 1966, 1967). In connection with studies on the natural ovipositing habits of female *C. p. fatigans*, especially in relation to septic tanks, a simple gauze trap has been designed to fit over the manhole of a tank. A louvered entrance a little less than a foot from the bottom of the cage allows access to gravid females attracted to the site. A gauze barrier at the bottom of the cage prevents the trapped females from passing through to the interior of the tank to lay their eggs.

As so much information about the behaviour and ecology of this important mosquito, with regard to blood feeding habits, infection rates and so on has in the past been based mainly on capture and examination of the house-frequenting population, there has long been a need for a fresh approach to many long-cherished ideas concerning this cosmopolitan vector of disease. There is already ample evidence that the Rangoon team is not only making striking advances in ecological knowledge on a broad front, but is also providing a new stimulus and incentive to further critical studies in other parts of the tropics.

In some regions the need for a wider sampling spectrum has been acutely felt, not only to deal with more critical standards required for the evaluation of filaria control measures, but also to deal with a possible range of culicine vectors differing in habit and availability. A good example of this development is provided by the filariasis research work carried out in Fiji, 1957 to 1959 (Symes, 1960; Burnett, 1960). The sampling methods used were as follows:–

(*a*) Daylight catches in houses with insecticide fog generators and groundsheets — usually between 8 a.m. and noon. This method was found to be twice as efficient as hand-catching.

(*b*) Bush catches — these were made by two men, one acting as bait for the other.

> i. A "standard" catch consisting of seven five-minute catches starting from a house at the edge of a village and proceeding in a straight line towards the catch points at 25, 50, 100, 150, 300 and 400 yd respectively.
> ii. A "track" catch started from the village and following a footpath with halts at the same distance as for the "standard" catch.
> iii. "Selected" catches were made in places offering shelter and likely to provide hiding places for adult mosquitoes, anywhere within about 500 yd of villages.

(*c*) Night catches: these were made by means of a small canvas hut fitted with window cages and adjustable shutters. Mosquitoes entered by way of the eaves, the shutters of which were closed during collection.

(*d*) Night-bait catch: one man (the bait) sat in his house with his shirt removed. The catcher was smeared with repellent. Catches were made for one hour at a time in different houses between 7 p.m. and 11 p.m. (e) All-day catches were also made by two different methods, the "track" catch and the "selected" catch repeated at two-hour intervals from 7 a.m. to 5 p.m.

An additional method was the carrying out of a comparison of night-bait collection and morning fog collection to see to what extent certain vectors left the house at or before dawn and thus escaped the fogging. With *Culex p. fatigans* the fog collection was much greater than the bait catch suggesting a build-up over the night with little exodus.

In the Malayan studies on *Mansonia* (Wharton, 1962) population changes were studied over a period of four years by trapping mosquitoes in stable traps baited with goats; this method was chosen as a result of trials which demonstrated that a goat attracts roughly the same numbers as man, and that a stable trap caught approximately the same number of the important *Mansonia dives/bonneae* group as did a human bait net trap.

Malayan *Mansonia* have long been recognized as exophilic, the great majority of those which feed indoors leaving for outdoor shelters at or before dawn. The search for this outdoor-resting population has been pursued very vigorously by a variety of methods. Apart, however, from limited success in kampong areas using box shelters, the searches in the forest for the elusive blood-fed population has been disappointingly unproductive. This has created a wide gap in the sampling spectrum not unlike that in the African studies on forest culicines (q.v.). It seems very likely that increasing light on the validity of different sampling methods for adult *Culex* and *Mansonia* will be thrown by the current quantitative approach to mosquito ecology. The need for more exact numerical data on the mortality of the aquatic stages — egg, larvae, and pupae — particularly with regard to mortality due to predators, has necessitated the development of emergence traps to record the day to day output of adult mosquitoes emerging naturally from breeding places (Laurence, 1966). It seems that progress in perfecting these emergence techniques could well point the way to a more ambitious attempt to estimate the output of adult mosquitoes from unit area in unit time, and thus give a better idea of the total or absolute population likely to exist in select experimental areas. Against this background it might be possible to view sampling data in a truer perspective, and obtain a clearer idea of exactly what proportion of the mosquito population is actually being tapped by different sampling methods.

The present approach to the ecology of culicine mosquitoes, particularly the vectors of filariasis, is being increasingly influenced by the development of new techniques for age grading females in the population (see also page 98). These techniques were originally worked out on anopheline mosquitoes by the Soviet workers (Detinova, 1962) but have since been applied to *Mansonia, Culex* and others (Bertram and Samarawickrema, 1958). These techniques are based on changes in the appearance of the female ovary associated with each gonotrophic cycle, i.e. the regular physiological cycle of feeding, ovarian development, ovipositing, and feeding once more. In its simplest form this age-grading technique enables a sample of the female population to be classified into nulliparous, i.e. those which have not yet completed one ovarian cycle; and parous, i.e. those which have completed one or more cycles. In its more advanced form, which is technically much more difficult and time-consuming, the bulk of the parous fraction of the female population can be further subdivided into those which have undergone 1, 2, 3, 4 or possibly more ovarian cycles.

The simpler technique distinguishing parous from nulliparous females makes it easier for larger samples to be dissected and has been widely used in mosquito studies. A great deal of that work has been concerned with investigating the natural mortality of mosquito populations using the principle that when the normal duration of the gonotrophic cycle has been determined (usually by direct observation on caged mosquitoes in the laboratory), estimates of the daily mortality of the population can be worked out mathematically from the parous rate (Laurence, 1963; Reuben, 1965). Both the simple and the more advanced age grading of female mosquitoes have also been employed to study the relationship between the age of the mosquito and its rate of infection with filarial parasites of different stages (Wharton, 1959; Samarawickrema, 1962, 1967), in much the same way as recent work on the relationship between age of tse-tse flies and their degree of infection with trypanosomes.

In the present context of the interrelations between problems of ecology and those of population sampling, some of the recent investigations on *Culex pipiens fatigans* in Burma and Ceylon are of particular significance. In Rangoon, Burma, simple age grading of females based on the proportion parous and nulliparous has been used to study changes in the population following control of breeding places (de Meillon and Khan, 1967). In one case successful control by drain cleaning was followed by a sharp fall in the density of adults, accompanied by an increase in the proportion parous, indicating that no newly emerged mosquitoes were being added to the population. In another instance a critical investigation by age-grading methods was made in connection with a sudden invasion of a particular culicine mosquito — *Aedes vexans* — recorded in the course of routine human bait captures. Both before and after this sharp invasion only negligible numbers of this species were found. Examination of the females caught during the peak showed that only 1% were parous, indicating that these mosquitoes had very likely all recently emerged about the same time from the same prolific breeding source.

In Colombo, Ceylon, the more advanced age-grading technique has been applied to the same mosquito — *Culex pipiens fatigans* — enabling the female population to be divided into five categories, viz. nulliparous up to 4-parous (Samarawickrema, 1967). This age grading was applied to samples obtained by three quite different capture techniques, namely, the house-resting population, the biting population, and the ovipositing population (gravid females caught as they alighted on the sides of open catch pits). In addition age grading was carried out on samples collected on human bait at different periods throughout the night. The results showed that the samples collected by the three different capture methods showed close similarity with regard to the age composition of the females. On the other hand the night collections on human bait disclosed some differences in the biting habits or biting activity of the females according to age. The nulliparous females, which formed 44% of the population, reached a peak of biting at 02.00 hours. The peak of the 1-parous was at 01.00 hours, while in the case of the 3 and 4 parous group (which formed 7·3% of the population) the peak of biting activity was reached earlier, between 22.00 hours and midnight.

GENERAL PROBLEMS OF MOSQUITO SAMPLING WITH LIGHT TRAPS

The use of light traps for sampling culicines has only been comparatively recently used in the African studies referred to (p. 55). However, the New Jersey light trap has long been a standard procedure in the United States, not only for the sampling of nuisance mosquitoes but also in studies of the culicine vectors of Western equine encephalitis, St. Louis encephalitis and others. There is an immense literature on this subject and it is only possible to select a few references which have a particular bearing on the sampling problems in general.

The systematic use of the light trap is illustrated by the Californian encephalitis surveillance programme (Loomis and Myers, 1960), which is designed to relate seasonal variations in mosquito vector density and incidence of human disease. This is mainly directed towards *Culex tarsalis*, vector of Western equine encephalitis, St. Louis encephalitis and others. In these extensive studies on *Culex tarsalis* in North America light-trap catches have been compared with other sampling methods such as mechanical sweep-net collections (Love and Smith, 1957), with artificial outdoor shelters (Loomis and Sherman, 1959), with man-biting after sunset (Beadle, 1959) and with collections from natural and artificial outdoor shelters, and traps baited with dry ice (Hayes *et al.* 1958). These comparative studies have revealed some very significant features about light trapping and its validity under different conditions. In the comparison between the light trap and the mechanical sweep-net, collections were carried out at weekly intervals throughout an 18-month period representing all seasons. The New Jersey traps were fixed at elevations of 6, 25, and 40 ft in a tower in a wooded area. The mechanical sweep-nets were operated at heights of 3, 6, 15, 25, 40, and 50 ft above the ground, rotating at about 25 revolutions per minute. The different kinds of traps were operated on different nights. An "index of attraction" of light traps for different species was obtained by dividing the total number of specimens caught in the light trap by the total number taken in the mechanical sweep-net. This index varied from 0·24 in the case of *Psorophora ferox*, relatively rare in light trap collections, to 19·04 with *Culiseta inornata* which was common in light traps but relatively rare in sweep-nets. The comparison revealed within these extremes a wide range of attraction according to different species. For example, the three most abundant species taken in light traps were *Uranotaenia sapphirina*, *Anopheles crucians* and *Aedes vexans*. On the other hand three species of *Culex*, including *C. quinquefasciatus*, all showed a very weak attraction to the light traps. Similar sampling limitations of light traps with regard to particular species were shown in the series of light-trap catches carried out in Puerto Rico in connection with the mosquito fauna of the international airport (Fox, 1958). These observations confirmed that the light trap failed to give an indication of the true abundance of *Culex quinquefasciatus*, the vector of filariasis, although it was common in the island. In comparisons of light-trap catches with those taken in artificial box shelters, it was found that *Culex tarsalis* was taken in about equal numbers by both methods, but that considerably larger numbers of *Anopheles freeborni* were taken in the box collections.

Within the same species it also appears that the attraction to light is by no means uniform throughout the life of the mosquito. Extensive studies on the dispersion of *Aedes taeniorhynchus* in Florida have defined the action of this sampling method more accurately by saying that "light traps sample that segment of the population which is on wing and responding positively to light" (Provost, 1952, 1957). These studies indicated that, with this species, the main migratory flight occurs before the females start the search for blood, and before they can react positively to light. There may be important phases in the life of the mosquito when the light trap provides no attraction. With this species at least there is considerable evidence that the attraction may by cyclical, females reacting positively to light on the seventh day of adult life, and about every fifth day afterwards (Nielsen and Nielsen, 1953).

An illuminating comparison between light-trap catches and two other sampling methods using bait was carried out on *Culex tritaeniorhyn-*

chus in the extensive ecological studies on Japanese B. encephalitis in Japan (Scherer and Buescher, 1959). In this case the light trap was used as a check on the more extensive bird-baited trapping which formed the routine sampling method. Both methods indicated highest density of *C. tritaeniorhynchus* in July, followed by a sharp decrease in early August. However, parallel collections from a near by pig-baited trap continued to be large throughout August and early September. The suggestion was put forward that below certain density levels pig bait was more sensitive than the bird-bait catch or the light-trap catch, and the authors point out that this illustrates once more the uncertainty accompanying the use of a single trapping method in attempting to evaluate total mosquito populations.

The New Jersey light trap was also used in studying the ecology of West Nile virus in Egypt (Hurlbut and Weitz, 1956; Taylor *et al.* 1956). These traps were operated from sunset to sunrise on four nights a week throughout the year, and provided the basis for seasonal records of culicine incidence month by month, and also for geographical distribution. Two of the main culicine species taken, *C. antennatus* and *C. univittatus*, were caught in large numbers and both showed a marked peak in August and September. The third main species, *Culex pipiens*, was also taken in the light trap, but at much lower densities and without showing a marked peak in the favourable summer months. This is apparently related to the fact that *C. pipiens* was the only species found in large numbers resting indoors by day — up to 515 in a single bedroom. A trial run with a man-baited trap operated simultaneously with a light trap showed that relatively large numbers of *C. pipiens* (and also *Anopheles pharoensis*) were taken in the man-baited trap. Bird-baited traps on the other hand caught mainly *C. univittatus*.

A critical examination of various factors influencing the efficiency of light traps has revealed additional unexpected sources of variation (Barr *et al.* 1960, 1963). For example, light traps operated within a few feet of each other still show wide variations in catch, especially at high mosquito densities. This phenomenon, somewhat similar to the "trap idiosyncrasy" revealed in studies on carrion-baited traps (Chapter 6), could be counteracted to a certain extent by having the traps constantly changing position. The variation between traps was also decreased by rotation, provided the average per trap was about ten mosquitoes; at low mosquito densities this seemed to have no effect. An additional complication was disclosed in a series of trials with unlighted traps in which the light bulbs had been removed, but fans were left operating. These unlighted traps proved to be attractive to some mosquitoes, e.g. *Culex*, almost as much as lighted traps. Perhaps the attraction to unlighted traps could be due to the fact that *Culex* in this case are really seeking a day-time shelter, perhaps at dawn, and are mechanically sucked in by the fans.

This critical series of trials also dealt with the question of light intensity, the traps being equipped with 25, 50, and 70 watt incandescent bulbs. The number of mosquitoes collected was directly related to the light intensity, although in the case of *Culex quinquefasciatus* the increase was not significant. There was no indication from these experiments that light intensity ever reaches the stage of becoming repellent. The opinion of the authors on another important aspect of light trapping is well worth quoting as it deals with a variable which influences all sampling methods based on the attraction of insects to light, to animal bait, to carrion or to other attractants, namely: "No way has been devised by the writers for determining from what distance mosquitoes are attracted to a light trap, but the results and observations recorded in the literature suggest it to be very short".

Some of the advantages, as well as the limitations of using light traps for sampling mosquito populations have been well demonstrated by the extensive survey carried out in Panama over a period of three years (Blanton *et al.* 1955). The standard New Jersey light trap proved to be by far the most effective method for establishing the distribution not only of the anopheline mosquitoes primarily planned for but also of a wide range of other insects of medical importance in the Canal Zone and in the interior of Panama. This survey, involving 138 different localities, was the most extensive light trap operation ever carried out in any country.

This investigation also confirmed one feature of light trap sampling which sometimes tends to be overlooked, namely, that predominantly diurnal species of mosquito will almost certainly be absent or under-represented in such light trap collections. This may be rather a serious deficiency in those cases — as in Panama — where the diurnal species include several of the culicine *Haemagogus* which play a vital role in the transmission of yellow fever virus.

Despite many of the uncertainties affecting the validity of sampling by means of light traps, the routine and standardized operation of this technique has on occasions been able to demonstrate dramatic changes in species composition, changes which almost certainly reflect on the epidemiology of the various vector-borne diseases concerned. For example, comparison of light trap catches made in south-western Georgia, U.S.A., in a year of normal rainfall, and in a year of low rainfall after several successive years of drought, showed that the three dominant species — *Anopheles quadrimaculatus*, *An. crucians* and *Uranotaenia sapphirina* — which composed 98% of the catch in normal years, were replaced almost completely in the years of drought by *Aedes* and *Psorophora* which now composed nearly 98% of the light trap catch (Smith and Love, 1956).

CHAPTER 6

BLACK-FLIES, SAND-FLIES AND MIDGES

It is convenient to consider these three groups of biting flies together. They are all rather small, elusive insects, which nevertheless can some-times occur as adults at such high population density that they con-stitute a serious biting problem. This applies particularly to black-flies, *Simulium* sp. and midges, *Culicoides* sp. which in fact have been almost as much studied as biting pests of man and livestock as in their capacity as vectors of disease. The available information on capturing and sampling these insects therefore comes from several different sources, often quite unrelated. It would accordingly be very instructive to see what bearing these different methods and experiences have on the many sampling problems which these three groups in particular possess in common.

It has been a very common experience with black-flies, *Simulium*, in all parts of the world, and particularly in those parts of Central America and Africa where *Simulium* is a vector of onchocerciasis, that, while females can readily be taken biting human bait outdoors by day, their activities and habits at other phases of adult life remain practically un-known. As far as tropical *Simuliidae* are concerned, therefore, sampling of the population has been almost entirely based on the catch of females attracted to bait (usually human) outdoors in the open.

Many investigators have looked for resting vector *Simulium*, but generally with disappointing results both in tropical Africa and Central America (Crisp, 1956; Dalmat, 1955). In West Africa good collections have on occasion been taken by sweeping vegetation, but the bulk of the flies taken have been females obviously attracted to the human collector. Males have been recorded in this way in small numbers, but as they can occasionally be taken in the human bait catch — although they do not suck blood — their presence may also have been determined by the human collector. Careful search of a wide variety of possible resting places of engorged or gravid females has been invariably negative (Crisp, 1956; Marr and Lewis, 1964; Lamontellerie, 1963; Le Berre, 1966). The actual technique of using human bait for sampling is usually based on an arbitrary exposure, during the day, of two or three bait boys, sitting with legs and ankles exposed, either collecting themselves or off each other. If few flies are taken after exposure of about 30 minutes, the bait moves a little distance and sits down again. Some collecting sites seem to be unduly favourable, while others, perhaps only 20 to 30 yards distant, are consistently poor. The interval between first exposure of the bait and the appearance of the first fly may also be very variable, and may involve a "waiting period" of several minutes even where man-biting *Simulium* is known to be very prevalent (Lewis, 1956). A rather more systematic development of this initial random collecting is the method used in Nigeria (Crosskey, 1958; Davies *et al.* 1962) of sampling by adult fly rounds. A series of catching points are arranged in a particular sequence and visited at regular intervals for a period of 15 minutes, during which time the fly boys catch all *Simulium* settling on them. By expressing these catches as "flies caught per boy-hour", density figures have been produced and have been used to record seasonal incidence of adult flies and changes in density before and after insecticide treatment of the rivers which form the breeding places.

It has also been recognized for some time that *Simulium* attracted to bait are not equally active at all times of the day. In the case of *S. neavei*, studies in Kenya showed a distinct biphasic cycle with peaks from 9 to 11.30 and again from 2 to 5 p.m. (McMahon *et al.* 1958). In routine surveys for this species the teams were usually in position by 9 a.m. the catching units taking up their positions, 50 yards or metres apart, in shady sites near the river's edge. The bait stood perfectly still while the searchers caught all flies. At the end of 15 minutes the team moved to a new site about 200 to 300 yards distant. A stretch of two to three miles of river was sampled each day in this way, and the density expressed as "flies per boy-hour". The differing degree of activity throughout the day was also found to be a feature influencing the availability of the three main species of *Simulium* concerned with the transmission of onchocerciasis in Guatemala (Dalmat, 1955). Although the biting activity of the three species covered the entire day-time period between the hours of 6 a.m. and 5.30 p.m. the following differences were revealed.

S. ochraceum: bites mainly from 7 a.m. to 4 p.m. but most active from 8 to 10 a.m.

S. metallicum: will continue feeding until 5.30 p.m.

S. callidum: prepared to bite from about dawn until about 9 a.m., and again from 3 to 4 p.m. until twilight, with very little activity in the middle part of the day.

In order to deal with such variables, a further refinement, used particularly with regard to *S. damnosum* in Africa, has been to carry out 12-hour catches from dawn to dusk using a team of collectors (Lewis *et al.* 1961). Under these conditions it was found that man-biting *Simulium* were usually more active in the morning and in the afternoon than around midday. By noting the number of nulliparous and parous in the catch, it has also been shown that in some areas the morning and afternoon peaks for parous females occur respectively earlier and later than those of the nulliparous (Lewis, 1960). In other areas, however, the same species — *Simulium damnosum* — has exhibited a rather different cycle with no morning peak or midday lull.

The extended 12-hour catch on human bait is also the main sampling method used by the French workers in West Africa (Le Berre *et al.* 1964; Le Berre, 1966; Ovazza *et al.* 1965) using teams of two bait/collectors, working in relays, who collect all the *Simulium* settling on them between 6 a.m. and 7 p.m. Collections are made once a week in this way in various localities in order to follow changes in adult density throughout the year in different environments. Latterly, the catches of female *Simulium* have been divided either into two — nulliparous and parous — or three physiological groups, namely, nulliparous, young parous, and old parous, a refinement which has proved of considerable value in trying to unravel some problems in *Simulium* ecology (Ovazza *et al.* 1965a) (see Fig. 6).

In these very empirical sampling methods with man-biting species it is usually assumed that the methods, despite their admitted imperfections,

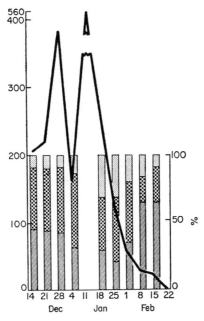

FIG. 6. Changes in the composition of the female population of *Simulium damnosum* captured on human bait in the Upper Volta region of W. Africa during the period in the dry season when flow in the main riverine breeding places is progressively decreasing until it ceases completely (after Ovazza *et al.* 1965:)

— Number of females captured;   ▨ %nulliparous;   ▦ %young parous;   ▧ %old parous;

are at least a measure of biting rate or biting activity. This assumption is probably well justified in the case of *S. damnosum* and some other species. However, some observations on temperate climate *Simulium* indicate that this assumption may require more critical examination. With *S. ornatum*, for example (Davies, 1957a, b), it was found that, of flies settling on cows, perhaps only a small proportion actually bit and that the time spent on the host animal may be greatly reduced when fly activity is high and when there are more flies settling on the cow in unit time. In the latter case there is an increase in the proportion of flies which manage to escape capture.

While progress in sampling the vector species of *Simulium* is obviously very slow in developing, experience with temperate climate *Simulium* has indicated possible lines of amplification. In studies on *S. ornatum* in Europe (Davies, 1957a), the population of gravid females has been sampled by the use of a type of sticky trap which was originally developed in connection with aphis studies (Broadbent, 1948; Broadbent *et al.* 1948; Broadbent and Heathcote, 1961). These traps consist of metal cylinders 15 cm in diameter and 30 cm long, covered with a detachable plastic sheet coated with commercial tree-banding grease. Traps suspended horizontally over a stream caught many females, all fully gravid, as well as a small number of males. In the same investigation no resting gravid females could be found.

Perhaps the most striking development in adult *Simulium* capture has been the extensive studies carried out in Scotland over a period of 51 months on black-fly populations taken in light traps (Davies and Williams, 1962; Williams, 1962, 1965). Although *Simulium* are normally considered diurnal insects, the success of these light-trap catches indicates an unexpectedly high degree of activity after dark. The procedure adopted in those studies was to use a light trap with a light source of ultra-violet and mixed wave-length light. The light was automatically switched on half an hour after sunset and switched off half an hour before sunrise. These studies produced an immense amount of data about the seasonal distribution of ten species of *Simulium*. From this wealth of information one or two points emerged which are of particular interest in the present context. Firstly, it was clear that this method of catching or sampling was not equally valid for all species. With *Prosimulium hirtipes*, for example, only seven specimens were taken in the five seasons although collections of aquatic stages showed that it was the most abundant species in the area. Other species, less common in breeding places, were consistently taken in great abundance.

The light-trap catches with *Simulium* in Scotland were particularly valuable in indicating seasonal distribution, adults of several species being taken over a much longer period of the year (including December in some cases) than had been previously considered on the basis of the occurence of aquatic stages. Both males and females were taken in these traps: the proportion varied considerably, the availability of the males being particularly subject to variation. It is still too early to judge the possibilities of developing light-trap methods of sampling tropical vectors of onchocerciasis. Trials with a very simple type of light trap in West Africa have shown that under certain conditions (page 75) some species of *Simulium*, both males and females, can be taken in this way, including a few *S. damnosum* (Lamontellerie, 1963, 1967). However, the more sophisticated light trap used in the Scottish studies described above has yet to be given a full and fair trial with vector species. Much depends on the essential point of whether these species are attracted to light in the first instance, and on the extent to which the quality or nature of the light source may be a factor determining success or otherwise. Certain modifications in basic design are indicated for the most efficient tropical use, and allowance would also have to be made for the possibility that trials might have to be done in areas where no mains power supply is available. Full details of such likely modifications, as well as recommendation with regard to all aspects of operation under tropical conditions have been provided by Davies (1965).

Should it be possible eventually to design light traps which prove attractive to *S. damnosum* and other vector species, many new possibilities might be opened up. Such a technique might well provide a more sensitive index of low adult vector populations, while at the same time sampling females of a wider range of physiological condition, perhaps including those with blood meals from sources other than man. There are already indications from the West African work referred to that the male populations might also be sampled in this way. This, or some similar method, would be of great advantage for sampling vector species in areas of their distribution where they do not normally come to human bait, e.g. *Simulium damnosum* in Rhodesia.

The whole question of *Simulium* sampling was discussed at a W.H.O. sponsored international meeting of *Simulium* workers in Geneva in 1964, (W.H.O. 1965) and it was again emphasized that the principal information about the density and distribution of *Simulium damnosum* and *S. neavei* was still dependent on the use of human catchers, in the form of "fly-recorders" or "vector-collectors", using themselves as bait. In view of the long recognized limitations in depending on this method of sampling alone (so liable to human error and to differences in attractiveness of the bait) it was considered essential that alternative sampling techniques be developed. Among the methods which appeared to merit further investigation with regard to the tropical vectors of human onchocerciasis, attention was drawn to the possibilities of the light trap (see above) which has not yet been given a full and fair trial with these species, and also to the use of chemical attractants (or extracts) as developed by the Canadian workers (Fallis and Smith, 1964). Another sampling method worth trial is the use of suction traps. These have already proved to be a useful additional method in the Scottish light-trap investigations described above (Williams, 1965). As this method has the advantage of being non-attractant, it might prove useful for dealing with populations which do not come readily either to bait or to light.

The urgent need to extend the sampling spectrum to the resting population of *Simulium*, particularly to the females during the stages of blood meal digestion and ovarian development, is now being increasingly appreciated, and clearly demands a new and vigorous approach. It is very easy to be discouraged by the fact that so far so little success has attended the painstaking efforts of so many competent Simulium workers, and that the chances of detecting such small insects in the mass of undergrowth, jungle-covered ravine, and tropical forest seems doomed to failure. If these day-time or night-time resting places are completely diffuse, without any tendency for concentration in a well-defined microhabitat, then the search will undoubtedly be a long and arduous one. But if, as seems perhaps more likely, the bulk of the resting takes place in comparatively restricted loci, then the prospects of detecting these eventually — perhaps by means of entirely new trapping devices — seems more hopeful. One particular line for investigating the specific problems of tropical vector species is suggested by observations made on the movements of adult Canadian *Simulium*, and the techniques developed in the course of that work (Wolfe and Peterson, 1960). In that investigation, *Simulium*, probably *S. venustum* for the most part, were actually observed moving to night-time resting places in the tree tops, and were again observed at dawn flying down from the tops of the forest cover. Following this up, observers climbed trees at night, shook the vegetation, and collected the disturbed black-flies by sweeping around with nets at 25–30 ft above ground. Whether any tropical species have similar resting habits, or whether resting is restricted to ground level, still remains obscure, but it is clear from the Canadian experience that the possibility of main resting places being in the trees or in the forest canopy cannot be overlooked, and that new and imaginative collecting techniques will almost certainly be required. In this connection much guidance and encouragement is provided by the great progress which has been made in recent years for studying day-time resting habits of tse-tse flies in Africa (Chapter 3).

The present reliance on a single sampling technique is well illustrated by investigations on two important aspects of *Simulium* ecology in West Africa. The first of these concerns the still controversial question of what happens to *Simulium damnosum* in savannah areas where the rivers which form the breeding places of this insect stop flowing and dry up during the long dry season. The early colonization of the rivers as soon as they begin to flow at the beginning of the rainy season, and the early appearance of adult female *Simulium* at that time, has been attributed either to the rapid reinvasion of savannah areas by flies from distant permanent breeding sites, or to the possibility of *Simulium* surviving the dry season in egg or adult stage. In studies in the Upper Volta (Ovazza *et al.* 1965b) it was noted that as soon as the rivers stop flowing, or only a very short time afterwards, female *Simulium* are no longer taken on human bait. Before this happens, the final collections are composed almost entirely of nulliparous females, and young parous, with the older parous group fading out (Fig. 6). As the population diminishes therefore, its physiological age also decreases. One explanation provided by the author for this seeming paradox is that at this time of the dry season there is an increased mortality, or decreased expectation of life, which prevents existing females from reaching an advanced parous stage. The other possible explanation considered by these workers is that at this time the older females are less attracted to man, and perhaps show an increased movement to resting places as yet undetected. As at present the only fraction of the *Simulium* population which can be sampled is that taken on human bait, it is evident that this fascinating ecological problem is likely to remain unsolved and speculative until such time as additional sampling techniques, not restricted to human bait, can be successfully developed.

How valuable such additional methods can be in certain situations has actually been demonstrated in that part of West Africa with another species of *Simulium*, *S. adersi*, which attacks man in parts of the Upper Volta region (Lamontellerie, 1963). By the use of a very simple design of light trap in the form of a 100 watt frosted electric light bulb placed over a white tray filled with water, both males and females of this species have been taken throughout the dry season in this savannah zone.

The second of these investigations on ecology of *Simulium damnosum* concerns the long-recognized differences between the epidemiology of onconcerciasis in the rain forest areas of West Africa, and the savannah zones exposed to a long dry season. Entomological investigations in the Ivory Coast and in the Upper Volta have again had to rely exclusively on human bait captures for sampling adult vector populations. In the rain forest area there is a high and more or less continuous production of *Simulium*, which disperses over a wide area. In the savannah areas there are wide seasonal differences in vector density, but a much closer concentration near the breeding streams and rivers and associated human settlements. Examination of bait-collected females has also shown that there is a progressive increase in the percentage parous when proceeding from forest zone to savannah (Le Berre *et al.* 1964; Le Berre, 1966). It is this high concentration of longer-lived females near the breeding areas which is considered to be the main reason for the increased risk of human infection in villages and settlements in those parts of the savannah.

Again, one feels that the two environments in which Simulium is being sampled are so markedly different that reliance on a single sampling technique is a severe restriction in such an ecological enquiry. Quite apart from climatic and topographical differences between rain forest and savannah, the question of alternative hosts in the rain forest area — normally thinly populated by humans — may be an important factor in determining the validity of sampling based on attraction to human bait. In addition there are obviously many pitfalls in interpreting age-grading data, based on human bait catch alone, in terms of the *Simulium* population as a whole in such forest areas.

Closely bound up with the problem of improved sampling for *Simulium* is the need for more accurate information about the feeding habits, especially of the vector species, and the range of animals attacked. This is particularly urgent in view of the likelihood that *Simulium damnosum* is composed of two or more subspecific groups, each perhaps attracted to man in a widely different degree (Dunbar, 1966; Duke *et al.* 1966; Lewis and Duke, 1966; McRae, 1967). Observations on Canadian *Simulium* have shown that with one particular species which normally feeds on birds, there is an increased tendency to feed on man with the onset of colder conditions in the autumn (de Foliart and Rao, 1965). Similarly, this species shows a change to man-feeding habits at higher altitudes.

These additional variables further underline the limitations of human bait sampling on its own. This may be specially pertinent at very low or negative densities where a false impression of vector scarcity or absence could easily be produced at a time when the vector was still present, but for some climatic or other reason was attracted temporarily to hosts other than man. The use of bait animals other than human has been used rather sporadically in the case of the *Simulium* vectors of onchocerciasis, and often under conditions in which results may be complicated by the presence of the human catcher, e.g. collecting on horse or cow. The possibility of developing more refined methods involving rodents or other small animals or birds as bait, uninfluenced by any human presence, is suggested by the following. — Methods worked out in the study of insect vectors of myxomatosis in Australia (Myers, 1956; Dyce and Lee, 1962), and which have successfully included *Simulium*, are based on a "cone-drop trap" which is suspended over a one-inch mesh cage containing the single bait (rabbit in this case). The cone traps can be raised and lowered by cords from a distance of 30 to 40 yards. The bait is exposed for 15 minutes at a time, after which the cone trap is lowered, and a collection is made of the insects trapped in the course of feeding on or attraction to the bait.

The same principle, with slight modifications, has been used in *Simulium* studies in North America, using various small animals and birds as bait (Anderson and de Foliart, 1961). In those investigations a dark box containing a small window trap of netting is lowered periodically over the exposed bait; insects attracted to the bait or feeding on it are attracted into the light cage which can then be isolated by a sliding panel to facilitate collecting.

In studies on *Simulium* attracted to various types of bird bait in Canada (Bennett, 1960) the same principle of bait exposure was used. The birds were placed in ordinary chicken-wire cages for 20–30 minutes at a time, after which they were covered by a fine mesh collecting cage for a further 20–30 minutes. This allowed engorged flies to leave the birds and settle on the walls of the cage where they could be collected by aspirator. By means of a simple rope and pulley system, the exposure cages containing birds could be raised to various heights up to 20 ft above the ground.

Other work in Canada on the attraction of *Simulium* to bird-bait has disclosed an unusual degree of specificity on the part of one species and has led to a discovery of considerable significance in vector ecology as a whole (Fallis and Smith, 1964) (Fallis, 1964). *Simulium euryadminiculum* is strongly attracted to the common loon (*Gavia immer*) as a host to such an extent that even a dead loon, or the study skin prepared from it, attracts hundreds of flies. These observations strongly suggested that attraction depended on a specific olfactory stimulus. This was tested by exposing a dead loon, dead black duck, dead mallard duck and both dead and living domestic duck near a lakeshore, and collecting the *Simulium* attracted. With this choice, *S. euryadminiculum* continued to be attracted in hundreds to the dead loon, to the almost complete exclusion of the other live and dead birds. By making extracts of different parts of the body with different solvents, and exposing on the lakeshore paper towels which had been soaked with these different suspensions and allowed to dry, a most spectacular reaction was produced. Within minutes, hundreds of black-flies swarmed over the paper towel on which an ether extract of the loon's tail had been poured. In contrast very few flies were attracted to towels containing extracts from other parts of the body. The authors of that work considered that the specific olfactory reaction was the primary stimulus in bringing the flies to the host, but that at close range, visual stimuli also appeared to be important, influencing the behaviour of the fly with regard to landing on certain parts of the body, or on raised objects associated with the attractive source.

As *Simulium* are almost entirely day-time biters, the question of visual attraction to the host has received a great deal of attention, particularly with regard to European and North American species. At the moment it is difficult to say how far these findings might apply to the tropical *Simulium* vectors of disease, or how far they will help to illuminate or improve sampling methods based primarily on host attraction. However, the possible significance of this work cannot be overlooked as it deals with aspects of activity and aggregation common to a wide range of other day-time biting insects, such as tse-tse flies and tabanids, as well as to those mosquitoes, such as species of *Aedes* and *Mansonia* which are day-time or crepuscular feeders.

In studies on *Simulium arcticum*, an important pest of livestock in Canada, an artificial animal-type trap, or silhouette trap, has been designed in order to provide a more satisfactory method for routine sampling of black-fly populations (Fredeen, 1961). This trap which has many features in common with the "animal trap" used in tse-tse studies (see p. 16) is a frame constructed in the form of a four-legged animal. The underside is left open, but direct sunlight is admitted to the dark interior only through an opening in the top. Over this slit is fitted a clear plastic cone terminating in a removable glass jar. Black-flies attracted to this silhouette trap are therefore mechanically trapped in much the same way as tse-tse. In experiments with three forms of trap, viz. cow silhouette, sheep silhouette and pyramid silhouette, the first was the most effective, confirming that the efficiency of such traps appears to be a direct function of surface area and size of opening, rather than shape. These traps caught most black-flies during a one or two-hour period before and after sunset, that is at a time when animal hosts are normally most severely attacked. It appears that this trap only samples the attacking population of black-flies, and from the strict sampling point of view is probably subject to many — but not all — of the variables attending the use of live animal or human bait. Unlike tse-tse flies in similar circumstances, the trap does not appear to attract females seeking shelter only, or females at intermediate stages of blood digestion.

The question of shape, colour and movement of such visual traps has been further examined in North American species by means of the "Manitoba fly trap" in modified form (Peschken and Thorsteinson, 1965), and in European species by means of a range of artificial animal and bird silhouette traps (Wenk and Schlorer, 1963). The methods used in applying the Manitoba fly trap to *Simulium* were esentially the same, on a smaller scale, as those originally developed with horse-flies (see page 17). Flies attracted to the trap were caught either in the non-return chamber, in which a bag of sodium cyanide was suspended, or on tanglefoot applied to the target. In this way it was possible to show that stationary cylinders or plaques were more attractive than the same targets suspended so as to rotate in air currents. By using two-dimensional shapes in various designs it was noted that solid patterns in the form of triangle, square or disc proved more attractive than broken patterns such as X, Y, or □. However, when three-dimensional models such as various forms of cube or cylinder were tested, there was apparently no discrimination.

In the studies on European species of *Simulium*, a range of artificial animal and bird silhouettes were ingeniously constructed in such a way that the actual silhouette of the each animal could be varied as to position of head, ears, tail and other projections. These observations showed that different species of *Simulium* tended to be attracted not only to different "hosts" but also to particular parts of the same host.

Of particular significance from the point of view of sampling fractions of the *Simulium* population other than those obviously attracted to bait, were the North American studies on artificial ovipositing sites in the form of stakes of different colours, placed either horizontally or upright in a *Simulium* breeding stream. The attraction of different individual stakes to ovipositing female *Simulium* was judged by removing the egg masses at intervals and weighing them. In this way it was shown that the differences between the mean number of eggs laid on stakes of different colours were highly significant, with the highest attraction at the yellow and green end of the spectrum, and the lowest at the red-black end. Studies on tropical vector species of *Simulium*, such as *S. damnosum* in West Africa, have shown that twigs and leaves suspended in contact with the turbulent water in select parts of streams and rivers can provide attractive sites for mass ovipositing by females (Muirhead-Thomson, 1956c; Marr, 1962). Possibly a further and more critical examination of this attraction may yet provide a valuable sampling method for the gravid fraction of the population, about which so little is known.

Many of the problems encountered with sampling black-flies (*Simulium*) apply equally well to midges (*Culicoides*) and sand-flies (*Phlebotomus*). *Culicoides* for example has received a great deal of attention not only as a vector of human disease (Filariasis due to *Acanthocheilonema perstans* in the Cameroons and the Niger Delta) and animal disease (African horse sickness, blue tongue of sheep) but also as a pest insect whose mass attacks on humans create a nuisance or even a major problem in many countries.

One of the commonest methods of sampling pest *Culicoides* in the American continent is a variation of the human bait catch, fly densities being recorded by "landing rates", i.e. the number of flies recorded on an exposed human arm during a short arbitrary period. When the length of this exposure period is standardized, e.g. two minutes, and the landing rates recorded at regular intervals throughout the day or the season, a comparison of midge abundance in different sites or localities can be made. Regular observations over a long period can enable estimates to be made of seasonal abundance (Jamnback and Watthews, 1963).

In studies on *Culicoides* in West Africa sampling by human bait/ collector has been used, with the refinement that collections of individual species of proved or suspect vectors were made during the period when the species was biting at a high or uniform rate. For example, the maximum biting period for four species were as follows (Nicholas *et al.* 1953):

> *C. grahami* (suspect vector)
> *C. inornatipennis* $\Big\}$ 5–6.30 p.m.

> *C. austeni* (proved vector)
> *C. fulvithorax* $\Big\}$ 7 p.m.–5 a.m. Night biters.

In general the use of human bait for sampling *Culicoides* has run into all the variables already dealt with, such as wind, numbers of hosts, exact location, time of day, etc. (Kettle, 1962). Of particular interest has been the use of several alternative methods of sampling *Culicoides*. In an investigation on midge control in Scotland the use of sticky traps was carefully examined. These took the form of black-painted cylinders which were grease-coated and supported on wooden stakes about 6 ft above the ground. Traps of this kind arranged in parallel rows, and in a line extending for 1 200 yd, were used for sampling and distribution studies (Kettle, 1960).

Sampling of the Scottish midges has now been extended to include suction traps in which a Vent-Axia fan sucks flying insects passively into a collecting tube, the catch being sorted automatically into hourly collections (Reuben, 1963). Traps were arranged with their mouths 6 ft, 3 ft, and 6 in respectively above the ground. As this is a passive sampling method which does not involve any bait or other attractant, it is considered to give an unbiased estimate of the number of insects in a constant volume of air in a definite period of time. (Taylor, 1951; Johnson and Taylor, 1955). However, this flying population is evidently not truly representative of the population as a whole. Males were rarely taken in the traps, and the female catch was composed almost entirely of those without blood in the gut, and at a very early stage of ovarian development.

Experience with the use of light traps in *Culicoides* studies has provided an interesting parallel to the *Simulium* studies described above in revealing selection by different species. In records from Porto Rico (Fox, 1953; Fox and Kohler, 1950) the figures showed that *Culicoides furens* — the common human pest species — made up only 30.5% of the total *Culicoides* catch in three localities, while *C. inamollai*, which had hitherto been regarded as an obscure species, made up 61% of the total. Light trap studies on *Culicoides sanguisuga* — an abundant pest midge or "punkie" in the

N.E. United States and S.E. Canada, have shown that although high catches of this species can be taken, the collections cannot be considered as representative of the population as a whole. Very few males were taken, only about 5 per 1 000 females. Of the female population itself only 2·8 per 1 000 were gravid, and 2·9 per 1 000 contained a recent identifiable blood meal. It is interesting to note that despite all the variable factors involved, the light trap counts extending over several months showed a pattern of seasonal abundance of this species very similar to that indicated by the entirely different sampling technique of recording standard landing rate counts on the exposed human arm (Jamnback and Watthews, 1963) (Fig. 7a, b).

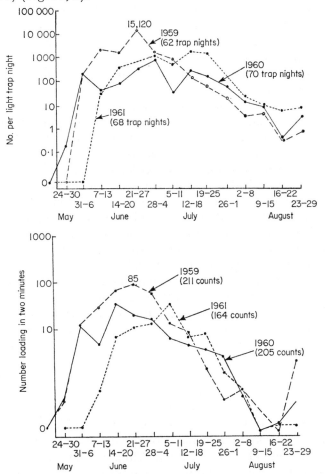

FIG. 7a and b. Comparison of two sampling methods for *Culicoides sanguisuga*, viz. female landing rates and female light-trap counts, during the same periods of the year in New York State (after Jamnback and Watthews, 1963).

These, and other factors, would clearly have to be taken into account in interpreting the results of another extensive light trap study carried out through a complete annual cycle, namely, the investigation on *Culicoides* populations in western Texas in connection with the epidemiology of blue tongue virus disease of sheep (Wirth and Bottimer, 1956).

As in other fields of vector ecology, there is an increasing need for non-attractant sampling methods for resting and in-flight populations of *Culicoides*. Among the methods which may well be improved and extended is the use of a power aspirator, and the use of large nets mounted on vehicles (Bidlingmayer, 1961; Nelson, 1966).

Many of the experiences gained, and the comments made, on the sampling of black-flies and midges, could apply equally well to another group of small elusive biting flies, namely sand-flies, *Phlebotomus*, vectors of human leishmaniasis. Early classical studies on the role of *Phlebotomus* as a vector of Kala Azar in India (Shortt, 1932) were concerned mainly with species which were abundant in and around human dwellings. Since that time investigations in other countries have given increasing attention to collecting or sampling *Phlebotomus* outdoors, both in the biting phase attacking humans or animal hosts, and in the resting phase in a wide range of outdoor shelters. (Kirk and Lewis, 1951; Petrischeva, 1946, 1965). The importance of the outdoor-resting population was early recognized by various workers, and Kirk and Lewis (1947) for example concluded that survival of sand-flies in some arid areas of Africa must be due in large measure to their exploiting a vast subterranean environment consisting of animal burrows, and deep cracks and fissures in the soil in which favourable conditions of temperature and humidity enabled them to avoid the severe conditions above the ground.

Sand-flies have always been a difficult group to sample, as distinct from just capturing. There are several methods of catching, collecting or trapping *Phlebotomus* and demonstrating apparent changes in species composition at different seasons, but it is still uncertain as to how much these captures, often restricted to specially favourable concentration sites, really reflect changes in the density of the population as a whole, or give anything approaching a true idea of vector distribution (Hoogstraal *et al.* 1962; Minter, 1964; Turner and Hoogstraal, 1965). In addition to the more or less conventional captures on human bait which are widely used, rodent-baited traps or cages are playing an increasing part in capture or sampling techniques for *Phlebotomus*, especially in epidemiological studies on the reservoir of leishmania infection in wild animals (Hoogstraal *et al.* 1962; Turner and Hoogstraal, 1965; Williams, 1965; Disney, 1966).

The actual design of sand-fly trap tends to vary according to the investigator, and there is already evidence that differences in the design, and operation of the trap, as well as in the nature of the host rodent, may make results difficult to interpret in some instances. For example, in the leishmaniasis studies in Sudan (Turner and Hoogstraal, 1965) it was found that *Phlebotomus langeroni orientalis*, a species which attacks humans in large numbers, and the suspected main vector of kala-azar in that area, was only taken in insignificant numbers in traps baited with four different species of forest-inhabiting rodents. In view of the fact that this species of *Phlebotomus* must almost certainly have other hosts besides man, the authors suspect that its reluctance to enter the traps must have been due to some structural feature, such as shape, size or aperture of entry cone — in much the same way as *Aedes africanus* in the forests of central Africa is reluctant to enter between the well spaced bars of a roofed cage baited with monkey, although it will readily bite monkey bait in the open. From data on those species of *Phlebotomus* which do readily enter such traps, and feed on the rodent bait, it also appears that quite apart from specific differences, larger sized rodents are more attractive than smaller ones. It is worth noting that attempts to counter this bias towards the larger bait animals have been made by adjusting the number of bait animals in each trap in order to balance the attractant mass. This has been done not only in *Phlebotomus* investigations (see below (Thatcher and Hertig, 1966)) but also in quite unrelated studies on feeding habits of mosquito vectors of Japanese B. encephalitis (see page 58).

Investigations in Panama have also yielded a great deal of valuable information on the use and limitations of traps baited with small animals (Thatcher and Hertig, 1966). A wide range of sampling methods was used in those studies, including a variety of baited traps, oiled papers alone or in combination with baited traps, and also collections in representative outdoor resting sites. Several types of baited trap proved to be unsuitable for various reasons, and the sampling method finally adopted was based on direct collection of sand-flies from caged bait. Various small bait animals were set up in cages in the forest in such a way that sand-flies have free unimpeded access to them. The traps were visited briefly for about 5 minutes every half hour, and collections made of all sand-flies settled on or near the bait. The collectors then withdrew to their base 50–100 yd distant in order to minimize the effect of any human attraction. These observations showed that the man-biting species of *Phlebotomus* in Panama have a wide host range among forest animals.

A combination of baited trap and sticky trap has been developed in leishmaniasis studies in British Honduras (Disney, 1966), and has been found particularly valuable in epidemiological investigations there (see page 84). This trap was designed in order to minimize the effect of sand-flies entering a baited trap and managing to escape before collection. A cubical rat cage is set in a tray surrounded by a flat horizontal surface which is coated with a film of castor oil, the whole being provided with a roof from which the cage and trap are suspended.

One of the more recent studies on the sampling of *Phlebotomus*, in the

Sudan, illustrates very well the careful approach to the elimination or selection of suitable sampling techniques (Quate, 1964). The oiled paper or sticky trap method was investigated, using 15 cm squares of wax paper coated with light oils such as castor oil and sesame seed oil, and suspended in a variety of habitats. With most local species the numbers caught were irrespective of whether castor oil or sesame oil was used, and if that had been the only consideration castor oil would probably have been adopted as standard because it lasts longer than sesame, which needs replenishing daily. However, for one of the most important species studied, *P. orientalis*, castor oil evidently had some repellence, larger catches being recorded on the sesame-treated papers. In the same way a similar type of bias or specific selection was revealed in experiments with light traps, which yielded mainly a less important species, *P. squamipleuris*, and failed to attract the others.

Observations on the use of human bait (or human "lures") confirmed how much this sampling can be affected by increasing wind movements, as follows:–

Below 1·5 m/sec	No effect
1·5 to 2·5 m/sec	Biting liable to fall off sharply
Over 4·0 m/sec	Biting ceases entirely

This author also draws attention to another factor which would appear to be equally applicable to a wide range of vectors in addition to *Phlebotomus*, namely the "unknown factor of population fluctuations of the mammals which serve as hosts for the seasonal sandflies".

At present, progress towards improved evaluation of sampling techniques for *Phlebotomus* is rather handicapped by the fact that in some of the main investigations, the exact role of the differing species of *Phlebotomus* as vectors of the local form of human leishmaniasis is still obscure, and it is not yet possible to concentrate studies on only one or two species. In addition, the epidemiological picture is further complicated by the need to study the vector in relation to possible reservoirs of infection in various animal hosts. In British Honduras for example, about fifteen different species of *Phlebotomus* have been taken in the course of three different capture techniques, namely, collections from outdoor resting places (mainly the buttresses of large forest trees), on human bait, and in traps baited with eight different species of rodents (Williams, 1965).

Similarly, investigations in the Gambia have employed a wide range of capture techniques, including sticky bands across burrows and other aggregation sites, box traps, captures at light, space spraying in houses, trap huts baited with man or goat and sticky traps near cattle (Lewis and Murphy, 1965). Again, a large number of species of *Phlebotomus* were recovered. This broad spectrum sampling is well in keeping with present needs and trends, and there seems little doubt that when studies can be concentrated on a few key species, light may well be thrown on many sampling problems of wide significance in vector ecology in general.

CHAPTER 7

HOUSE-FLIES AND BLOW-FLIES

All the groups of Diptera discussed so far can be accurately designated as vectors of disease. In most cases the disease-producing parasites
or pathogens which they acquire in the process of feeding on human or
animal blood undergo development and multiply in the insect body.
After the appropriate interval infective forms of the parasite are produced which are capable of being conveyed to a fresh host in the course
of biting or blood-feeding. With some biting insects, e.g. *Tabanus*,
Stomoxys, the disease organism may be conveyed directly from host to
host by simple inoculation through contaminated mouthparts.

The common house-fly, *Musca domestica*, and its other non-bloodsucking
allies, comes in rather a different category, namely that of a passive
conveyor of disease organisms from excreta or other contaminated
sources, to foodstuffs, utensils, and other components of man's immediate
environment where infection has every chance of being picked up by a
new host. The filth-feeding habits of blow-flies on carrion (*Calliphora*,
Lucilia, etc.) or on excreta (*Chrysomyia*), and their close association
with man and his habitations, also expose them to suspicion, although
their role as agents of human disease still needs clarification (Greenberg,
1964).

However, in the general context of sampling populations of winged
vectors of disease, the present review would be very incomplete without
examining the peculiar difficulties presented by house-fly and blow-fly
populations. In the same way critical studies on blow-fly populations,
connected with animal health rather than human disease, have proved to
have much in common with similar problems encountered in tse-tse work.

As the non-bloodsucking house-fly is not associated with any specific
"host", sampling methods have in general been based on estimates or
counts of flies in known concentration or aggregation sites, usually in and
around human habitations. The success of these methods depends on a
considerable knowledge of the range and type of concentration site
preferred by the fly. These sites may vary according to temperature, to
the time of the day or season of the year, and may shift from indoors to
outdoors depending on conditions. In addition to the more obvious
physical factors such as light, shade, temperature, etc. the degree of fly
concentration at a particular site may be influenced by some attractive

substance, such as urine, faecal matter, or offal, perhaps transient and difficult to define.

In an attempt at numerical accuracy one of the most widely used sampling techniques for house-flies is the count based on a standard grill or grid placed over known fly concentration sites (Scudder, 1947; Holway *et al.* 1951; Madwar and Zahar, 1951; Schoof, 1955; Hafez and Attia, 1958). This usually takes the form of a grill or framework 36 in square, with twenty four $\frac{3}{4}$-in wooden strips fixed $\frac{3}{4}$ in apart. Densities of flies are based on grill counts made at standard times or intervals.

The method appears to be subject to many human errors and unconscious bias. In practice, it appears that in many cases the numerical data provided by the grill count can hardly claim to be more accurate than visual estimates of fly density made by an experienced operator. (Welch and Schoof) 1953. An inherent variable of the grid count is well revealed by Schoof's observations in which he compared grid counts in the same concentration site, using grids of two sizes, viz. the standard 36 in size with 24 slats, and an 18 in grid with 12 slats. At low fly densities when the number per grid count ranged from 0 to 5, there was no difference between the two sizes of grid. At 6 to 25 flies per grid, the count in the 36 in grid was slightly higher. At high counts, 51 to 150 flies, the 36 in square grid recorded twice as many flies as the 18 in grid (see Fig. 8).

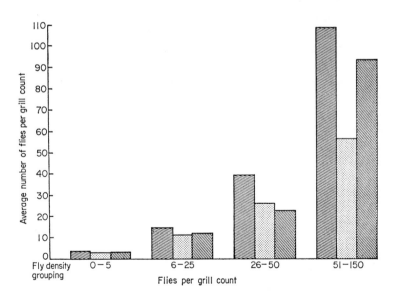

FIG. 8. Average fly densities in areas with differing fly potentials, as determined by three methods of surveillance (after Schoof, 1955):

Scudder 36 in square grill;   18 in square grill;   Reconnaissance.

Another variable of the grid count, which has to be taken into account when more than one species of fly is involved is the fact that the pre-ferred concentration sites of different species do not necessarily coincide, and that their proportions may also vary according to the nature of the attractant present. Interpretation of grill observations should also allow for the fact that two distinct reactions play a part in aggregation, viz. the aggregation due to more flies approaching the site, and aggregation due to flies spending more time on the site (Arevad, 1965), and that both reactions may be involved. The strict analysis of attraction sites and attractants used as a basis for sampling should therefore take account of the number of flies arriving in unit time, and the duration of these visits. In addition, quite apart from any chemical attraction causing fly aggre-gation on sampling sites, the physical nature of the substrate is also a pertinent factor, flies for example usually settle more readily on very rough surfaces such as straw and sacking than on smooth ones such as wood or concrete.

What might be regarded as a refinement or a modification of the grid count is provided by the technique of collection by vacuum or power operated aspirator (Anderson and Poorbaugh, 1964b) which was devel-oped specially to capture flies on the accumulation of droppings in poultry ranches. Despite the advantages in speed and efficiency, and the ability to cope with unusually high densities of flies, this method too has its limitations, in that it is not continuous but is confined to a single point in time.

Recently, increasing attention has been given to fly sampling methods which trap continuously over a period of time and which thus minimize the effect of the considerable short-term fluctuations in numbers at different times of the day (Raybould, 1964; Anderson and Poorbaugh, 1964a). One of the most promising of these methods is the use of the sticky trap which may be regarded as a scientific refinement of the old "tanglefoot" fly-paper method. Critical studies in East Africa showed that sticky traps made of long thin strips, with a proportionately smaller surface area, were much more attractive than the conventional form of tanglefoot fly-paper. In addition, improved adhesives have been devel-oped which retain the stickiness for a much longer period, without deterioration. By releasing known numbers of flies into an empty room containing fly-papers of various shapes and sizes, the relative attraction of different lengths and widths of sticky strips could be studied, as well as the patterns of location, height or level, in the rooms. In comparing the results of sticky traps with those of the conventional Scudder grill, it was found that the sticky traps gave figures of greater consistency than those of the grill, and provided a more accurate indication of weekly fluctuations in fly density (Raybould, 1964).

The relative values of these two entirely different types of fly-sampling technique — the sticky trap and the conventional grill — have been the subject of a particularly illuminating investigation dealing with several species of domestic and peri-domestic fly in East Africa, both indoors and outdoors (Raybould, 1966a, b). For this purpose, catching station of high, medium, and low fly density were selected, and the two methods of sampling were used alternately for weekly periods. During the week in which traps were being used in one group of stations, grill counts were made in a second group, the position being reversed in the following week.

Each fly trap consisted of a strip of polythene, 75 cm x 2 cm, weighted at the base and painted with an adhesive called "permanent tacky adhesive" (P.T.A.) whose ideal composition had been worked out in the preliminary tests described above. In each catching station a single sticky trap was suspended from the ceiling of the room in question. In low-density stations traps were left in position for one week, but where higher fly densities existed the traps were replaced before they became smothered with flies. The grill counts were taken four times a week, at fixed times in the morning and in the afternoon. A 3-ft square grill was placed on a high concentration focus, and the number of flies were counted at the end of a 30-minute exposure period. The results with reference to *Musca domestica* are summarized in Table X, and show that in practically all stations the trap index greatly exceeds the grill index, this superiority being in general particularly marked at high fly densities.

It appears that the main advantage of the sticky trap under those conditions is that it is continuous and cumulative, as compared with the grill count which is confined to a limited concentrations site for a relatively brief period of assessment during the hours of daylight. The grill count is much more liable to be influenced by short term irregular fluctuations in fly density during the course of the day, as well as by the more regular diurnal changes in which for example the afternoon grill count is usually higher than the morning count on the same day. The close dependence of the grill count on the nature of the attractant at the concentration site sampled became very obvious in one of the catching stations sited near a butchery — Group B, 1 in Table X — where the fly population fluctuated considerably according to whether meat was exposed for sale or not. The limitations of the grill count also became apparent in sampling domestic blow-flies such as *Chrysomyia*, which tends to disperse when disturbed by the grill placed on a concentration site.

In those East African studies an additional technique was used to extend the information provided by the normal grill count. This was the "adhesive grill count" in which a normal Scudder grill was coated

TABLE X

Comparison of fly-trap and grill methods for sampling populations
of *Musca domestica* in Tanzania, East Africa, at different population levels
(after Raybould, 1966a).

	Catching Station	Fly-trap index	Grill Index	Ratio Fly-trap/ grill
Group A.	1	2 020·8	8·9	227 : 1
	2	435·1	2·7	160 : 1
	3	441·2	4·4	100 : 1
	4	137·2	2·2	62 : 1
	5	59·0	1·3	45 : 1
Group B.	1	1 260·3	33·5	38 : 1
	2	382·0	1·0	382 : 1
	3	98·9	2·0	49 : 1
	4	37·5	1·8	21 : 1
	5	0·1	0·1	1 : 1

with adhesive enabling settling flies to be caught, and after being soaked off with kerosene, examined individually. In this way a more accurate idea could be obtained of the proportion of sexes included in the normal grill count. The relative proportions of males and females sampled by grill and by trap differ significantly in the two methods. Perhaps because female house-flies are more susceptible to food attraction than the males, the Scudder grill usually records a higher proportion of females than the non-attractant sticky traps. In this particular investigation the use of the adhesive grill count had the additional advantage of enabling closely related species of *Musca* — difficult to distinguish visually when settled — to be accurately identified.

Even the most domestic flies spend some part of their lives outdoors as well as indoors. In addition some peri-domestic flies associated with man are mainly outdoor species which do not use aggregation sites to the same extent as *Musca domestica*. In order to sample this outdoor population, the possibility of the sticky trap technique for outdoor use has also been closely examined. For this purpose the composition of the adhesive which had been found ideal for indoor purposes, had to be modified in order to make it stand up to the range of weather conditions existing in that area. In addition, a suitable form of upright trap had to be designed to replace the suspended trap which had been used in indoor studies. The observations showed again that the trap samples were consistently higher than the grill counts. At the same time, however, the traps evidently sampled a smaller proportion of the total population than they did indoors, probably due to the greater number of alternative resting places outdoors.

In appraising the value of different sampling methods there are great advantages in being able to work with known populations whose size and composition could be accurately controlled. In the case of the first series of East African experiments recorded above, this control took the form of releasing 200 house-flies at a time into an unfurnished room in which different types of sticky trap were suspended. With such insects, which normally spend so much time within such comparatively circumscribed spaces, there is nothing unduly unnatural or artificial about this controlled release experiment. The results can be even more illuminating when such experiments are designed to cover a range of population densities, as illustrated by the following example.

In Savannah, Georgia (Schoof, 1951) experiments were designed to evaluate the efficiency of the Scudder Grill technique at different known fly densities, the observations being made in an experimental cage 80 ft × 30 ft × 10 ft high, lined with plastic screen and set up in an open field. Four population levels of flies were used, viz. 200, 800, 3200 and 12000, the observations involving two species of flesh-fly as well as the house-fly, *Musca*. The flies used were all 3-day old adults raised by standard insectary methods. In addition to the basic experiments involving different population levels, and using different proportions of the three fly species, the influence of other controllable factors introduced into the cage — such as bait (milk-malt and standard garbage bait) and the presence of scrubs and shelter was studied.

These observations revealed that under all conditions the average grill count per run showed progressive increase with population increase: at the same time the relative proportion of the total population counted decreased with population increase. When garbage was present in the cage the average grill count was nearly doubled as compared with the cage without garbage, and under those conditions the difference between the percentage counted on the grill at high and at low levels was even more marked. For example, at a population of 200 flies in the cage, the average grill count was 27 (13·5%); at a population of 12800, the average grill count was 758 (or 5·9%). Another difference revealed was that the counts at the 200 population level were quite closely grouped, but the range of counts showed increasing variation at the higher population levels.

Further conclusions drawn were that the grill counts were influenced

markedly by species composition of the fly population; that the inter-
action of the cage conditions and the species composition also influenced
grill counts, especially at the higher population levels, and that there
was no significant difference between the grill counts made by the two
observers in this series of experiments. Those observers pointed out that
while certain major factors of possible influence on the validity of grill
counts could be controlled and reproduced accurately in these cage tests,
there were other uncontrollable factors which also had an influence on
the results. Such factors as time of day, temperature, wind direction,
relative humidity, light intensity, sky cover, etc. The analysis of these
uncontrollable factors, many of them not independent of each other, is
a complex problem, but at least the data showed fairly clearly that time
of day has a definite influence, and that there was a trend towards higher
grill counts in the afternoon. In addition, temperature had a marked
effect on the grill count with a fall at all population levels with falling
temperature below about 70°F.

As there are many factors which may play varying roles in determining the attraction of concentration sites where fly counts are likely to be
made, it seems that this sampling method will remain largely empirical
until progress has been made in understanding and unravelling these
factors. Two developments may illustrate the way in which behaviour
studies will assist in achieving this object. A great deal of attention is now
being given to the possibilities of house-fly control by insecticides and by
chemosterilants. The most convenient way of ensuring that these
chemicals are ingested by a high proportion of the fly population is to
incorporate them in especially attractive baits (Rosen and Gratz, 1959).
The first problem therefore is to re-examine the whole question of bait
attraction (Dame and Fye, 1964), and it seems very likely that one of
the indirect results of that work will be to place house-fly sampling on a
sounder scientific basis. Of particular interest is the fact that the chemosterilant studies lay special emphasis on the feeding habits of males, as
distinct from females, and that this refers particularly to the requirements
of the young recently-emerged males whose pre-mating intake of
chemosterilant is a vital factor in determining the success or otherwise
of this method of vector control.

The second type of investigation which may help towards a better
understanding of the factors involved in sampling is an analysis of house-
fly behaviour with regard to different objects in the environment (Mourier, 1965). Evidently house-flies show special behaviour towards "new
objects" introduced into familiar surroundings. Initially, such objects
prove highly attractive, but the attraction soon decreases, after perhaps
15 to 20 minutes, and the new object ceases to receive any preferential
treatment. Equally important is the fact that the intensity of this attrac-

tion to the new object is influenced by the physiological state of the fly, flies deprived of food and/or water showing a specially sharp reaction. As the introduction of a fly-grid into the fly's environment would appear to come in the same category of "new objects", it would be interesting to see if fly counts on such grills show a similar fall-off after the initial sharp attraction.

It is possible that this type of sharp reaction to novel stimuli enters into other aspects of insect sampling. Personal experience with wood-land *Aedes* mosquitoes in England (Muirhead-Thomson, 1956a) showed consistently that the number of *Aedes cantans* attracted to human bait entering the wood by day rose quickly to a sharp peak then de-clined after 5 or 10 minutes. When the bait moved only a few yards away to a new stance, a renewed wave of biting took place, and was followed by a decline. Finally the same phenomenon occurred when the bait moved back a few yards to the original stance. In the case of the re-activating of *Aedes*, stimuli may have been provided visually by the movement of the human "lure" from stance to stance, or it could simply have been the renewed presence of the human host which provided the fresh attraction of a "new object". Whatever the explanation, it is clear that one more factor must be added to the complex to which even the most simple and direct sampling techniques are exposed.

The use of baited traps for sampling populations of house-flies and blow-flies has also been attended by many difficulties and variables, particularly with regard to obtaining a bait of uniform attraction (Davidson, 1962; Norris, 1965, 1966). Some of the most searching in-vestigations of bait sampling have been done on blow-flies and carrion flies of veterinary importance (MacLeod and Donelly, 1957a, b; MacLeod 1958). As many of those species studied have also been investigated with regard to their possible role in the epidemiology of poliomyelitis, these findings are of added significance in the strict context of sampling insect populations of medical importance (Lindsay and Scudder, 1956; Nuorteva, 1959).

As the fly population of carrion-infesting calliphorines exists in a state of dispersion and low density, it is necessary to use an attractant which will concentrate the majority of flies within this influence. The obvious choice of attractant is carrion. However, it was shown that even a uniform type of bait material, such as rabbit flesh, does not exert a uniform attraction. This attraction varies according to the stage of dry-ing or decomposition, which in turn is influenced by hot or cool weather. Different species of carrion fly tend to be attracted at different times, and even within a given species the responses of the two sexes to the same bait may vary. The effective area of attraction of bait is also sub-ject to considerable variations, and is particularly liable to be drastically limited by the presence of other, perhaps unknown, attractants in the vicinity.

One of the most intangible variables is due to the trap itself, and has been designated "trap idiosyncrasy". Other factors being equal, certain traps prove to be consistently better, or worse, than the majority. As it has been observed that trapped blow-flies attract other blow-flies to the trap, it would appear that if one trap initially catches more than another — perhaps due to trap idiosyncrasy — the differences between the two might increase rather than decrease. In an efficient trap, numbers in excess of a thousand may be caught in the matter of an hour. This high catch itself may prove a limiting factor, in that a point may be reached when the high number of trapped flies appreciably diminishes the light intensity, with the result that there will be an increasing number of escapes to balance the increasing number of entrants.

In order to deal with these difficulties it would be necessary to:

(*a*) curtail the exposure period (not usually practicable);
(*b*) reduce the attraction of the bait;
(*c*) reduce the efficiency of the trap.

The only way to measure the bias which must exist in this type of sampling is to estimate the absolute density of the bait-trap sampled population, by means of the release-recapture method discussed below.

A different principle of sampling has been used in these same studies on blow-flies of veterinary importance. In that method, unbaited tent traps are used consisting of tents with a base of one square metre topped by gauze cones terminating in traps. The traps are put out at night resting on the ground in such a way that flies resting in the undergrowth are attracted by the morning light shining into the apical traps. It is considered that this method is probably the one most likely to lead to an unbiased sample of the population, provided it could be done on a sufficiently comprehensive scale to allow for the low density and irregular distribution of resting blow-flies.

CHAPTER 8

VECTOR POPULATION STUDIES BY MEANS OF MARKING-RELEASE-RECAPTURE

The marking, release and subsequent recapture of marked insects is a technique which has been extensively used for many years in studies on vector ecology. The bulk of those studies have been mainly concerned with problems of flight range, dispersion and distribution of adults, and cover a wide range of insects, such as house-flies (Quarterman *et al.* 1954; Pickens *et al.* 1967) mosquitoes (Gillies, 1961; Pausch and Provost, 1965; Lindquist *et al.* 1967) black-flies (Fredeen *et al.* 1953; Bennett, 1963) and tabanids (Beesley and Crewe, 1963). Any attempt to summarize or assess the mass of information resulting from all that work would be rather beyond the scope of this book, and accordingly attention will be concentrated on two particular aspects. Firstly, the use of the marking-release-recapture technique in population studies aimed at estimating the absolute density of vectors in a particular area, and secondly, the use of the technique in order to establish the relation between the physiological age of female vectors according to the condition of the ovaries, and the real or calendar age, in days.

POPULATION STUDIES AIMED AT ESTIMATING THE ABSOLUTE DENSITY OF VECTORS

When applied to population studies the principle involved in this approach is the Lincoln or Peterson Index which states that in a select study area in which a number of individual insects or animals have been captured, marked and released, and in which a subsequent catch in that area includes a certain number of the original marked individuals, the population in that area will be

$$\text{Total individuals captured in second catch} \quad \times \quad \frac{\text{Number of individuals marked and released}}{\text{Number of marked individuals taken in the second catch.}}$$

Although the method has been widely used in general ecological studies of insects and animals, the most significant studies from our point of view are those carried out on tse-tse in East Africa over many years

(Jackson, 1944, 1947, 1953,), and the later studies on blow-fly populations in England (MacLeod and Donelly, 1957; MacLeod, 1958). In the classical tse-tse studies nine different, coloured, artist oil paints were used. Each tse-tse was marked with two different colours at a time, 14 different spots being available for marking on the dorsal surface of the thorax. In this way it was possible to mark upward of 25 000 individuals without duplication, thus avoiding the need for remarking on recapture and further release.

In this work two different methods of calculating tse-tse populations were used:–

(a) finding the proportion of marked to unmarked flies at regular intervals after release, and extrapolating the proportion of marked to unmarked flies at the moment of marking (Jackson's "positive method");

(b) marking and releasing on a number of occasions, and then recapturing once (Jackson's "negative method").

In the more recent work on blow-flies a variety of marking methods have been used:–

(i) hand-marking of individuals with nitrocellulose paints incorporating light-resistant pigments, and using at least 10 positions for marking on the thorax of the fly:

(ii) mass radio-active tagging in which the flies are made radio-active by adding p–32 to drinking water, labelled flies being detected by screening:

(iii) mass powdering, in which the merest trace of coloured powder on a recaptured fly can be detected by means of a few drops of acetone which dissolves the powder and stains blotting paper:

(iv) powdering of radio-active tagged flies, the object of this being to facilitate detection of powder in recaptured flies by limiting testing to flies already rapidly picked out by radio-activity;

(v) self-marking with fluorescent material, this is an ingenious method whereby the movement of the fly's ptilinum in the process of forcing its way through the soil from a buried pupa, ensures that the fluorescent material mixed in the soil and sand is adequately picked up, and successfully retained and protected by the subsequent retraction or inversion of the ptilinum. This method has also been used more recently with tse-tse flies in Southern Rhodesia (Dame et al. 1965).

It will be noticed that the two insects which have been used successfully in these marking and recapture studies are not only sufficiently large to allow individual marking on the thorax, but sufficiently robust to stand up to the handling. Mosquitoes in general are much less suitable subjects for marking, and although individual hand-marking has been used successfully on a small scale in order to illuminate certain features of their biology, in general they can be more suitably mass-marked by dusting, staining, or by radio-active tagging in adult or larval form.

Both in tse-tse studies and blow-fly studies the proportion of marked individuals recovered has been sufficiently high to enable calculations to be made on the basis of the Peterson Index. However, the straight-forward use of this index is only justified if the capture follows immediately on the release of the marked flies. As it is, a certain time must be allowed for recovery of marked flies and for their mixing with the general unmarked population. Both in the "positive" and "negative" methods, therefore, the actual calculation is rather involved, and it would be out of place in this general review to examine it in detail. The reader will find an excellent examination of the whole calculation in Andrewartha (1961), Parr (1965), and Southwood (1966).

Those who have been concerned with these population studies are fully aware of the many snags and obstacles in the way of reaching a correct interpretation. As mentioned above, a certain interval must be allowed after release of marked individuals in the general population in order to allow free dispersal among the unmarked population before a second capture. Our calculations assume that this dispersal will be homogeneous, but that may not be so. In addition, the very interval allowed for mixing also permits dispersal beyond the collecting or study area, and it also permits the mortality factor to intervene. We have to assume that over a short period of days at least there is no significant difference in mortality between marked and unmarked individuals. The method of releasing marked individuals is also highly important. The method of catching wild flies, marking them and releasing them on the site of capture seems to deal with many objections, and would appear to be superior to the use of large numbers of marked individuals — perhaps bred out in the laboratory — being released from a central point. Certain methods of release may also accentuate the natural escape reaction of freshly-released flies, possibly producing over-dispersion beyond recapture limits. In the case of savannah tse-tse, for example, one can imagine that if sampling is based on the fly round the very flies liable to be caught for marking — mostly males — might by those individuals most likely to be recaptured on subsequent catches by the same technique. In this way an artificially high recapture rate would grossly underestimate the real population.

In the blow-fly studies, baited traps were used as the sampling method of choice in these recapture studies. Here the uncertainty arises as to the exact extent of the area or zone from which flies are attracted to one trap. In addition the attraction to the bait trap may vary from species to species, or between the different sexes of the same species, and possibly at different physiological stages (Norris, 1966).

A similar source of bias in recapture was revealed in marking-release experiments with house-flies in the southern United States (Murrosh and Thaggard, 1966), in which fly-ribbons (sticky fly-papers) were used to capture flies. When these fly-ribbons were put up on the same day on which batches of marked flies were released, it was noted that an un-expectedly high number of marked flies were caught on the paper. This was found to be due to the fact that young flies (which make up the bulk of the release batch) have a natural tendency to fly up to a greater extent than older flies, and were consequently more liable to be caught on the fly-ribbons. This particular obstacle was countered by not putting up the fly-ribbons until the day after release.

An important factor in the analysis of capture-recapture data in estimating daily populations has emerged from work on dragonflies (Parr, 1965), in which it was concluded that some methods of analysis are likely to be more effective when low recapture data are obtained from large populations, while other methods of analysis work best when high recapture data are obtained from a relatively small population.

AGE-GRADING OF MOSQUITO AND TSE-TSE POPULATIONS

In the past few years there have been considerable advances in knowledge about the changes produced in the ovaries of female blood-sucking flies as a result of the regular cycle of feeding, development of ovaries, egg-laying (or depositing of larvae in the case of tse-tse), and blood-feeding once more. This knowledge has been applied to the development of methods for estimating the physiological age of in-dividual females, and for age-grading natural populations. In the simpler form of age grading, which has been widely used in studies on mosquitoes and black-flies, a distinction is made between the two main categories of nulliparous, i.e. those which have not yet completed one ovarian cycle, and parous, those which have completed one or more cycles.

In the more advanced forms, the age grading techniques enable further distinctions to be made within the parous fraction, according to whether females have undergone 1, 2, 3, 4 or possibly more complete ovarian cycles. The application of these advanced age-grading methods to problems of vector ecology has been particularly marked in the case of anopheline mosquitoes and tse-tse flies. As the length of each cycle from oviposition to oviposition (or larviposition to larviposition in the case of tse-tse) can be assessed fairly accurately by observations on females in captivity, in theory it should be possible to express different physiological age categories in terms of real or calendar age. In the case of many tropical species of anopheline mosquito for example, this cycle normally covers 3 to 4 days under favourable conditions. In the case of tse-tse flies, the cycles between each deposition of larvae are of rather longer duration, say 11 to 12 days.

However, it would clearly be more satisfactory to obtain something more direct and conclusive about this important relationship between physiological age and calendar age. It cannot always be assumed that in nature the cycles conform to a completely regular pattern. Adverse climatic conditions, or obstacles in the way of obtaining a blood meal or in finding a suitable site for laying eggs or depositing larvae, might prolong the cycle and upset calculations. The marking-release-recapture technique seems to offer unusual scope for testing this question in a direct manner.

In the mosquito field, the most illuminating work on this problem has been carried out on anopheline mosquitoes (*A. gambiae* and *A. funestus*) in Tanzania (Gillies and Wilkes, 1963, 1965). Early attempts to apply this specialized technique of advanced age grading to these tropical vectors of malaria had been rather discouraging, but eventually the difficult dissecting technique was mastered, and the age-grading method was shown to apply both the *Anopheles gambiae* and *A. funestus* (Detinova and Gillies, 1964). This could be confirmed by dissections of laboratory-bred females which had laid known numbers of egg batches up to fourteen.

In order to establish the relationship between physiological age and real or calendar age, just over 10 000 female mosquitoes which had emerged in the laboratory — mainly throughout the hot season of 1964 — were marked by topical application of coloured paints (Gillies, 1961) and released on seventy-five separate occasions, mostly in batches of 100 to 200. Sixty-one of these marked females were recaptured after various intervals, and all but one dissected for physiological age determination. The results are shown in Table XI.

Although half of the recaptures were nulliparous or pre-gravid, sufficient numbers of older females were caught to give a very good indication of the relationship. Of particular interest is the female recaptured thirty-four days after release, which was found to have laid ten times, and also to have sporozoites in its salivary glands. Earlier in this same investigation, age grading determination had been carried out on a wild population of *A. gambiae* and *A. funestus* captured in houses, and in the course of that work a single 12-parous *A. gambiae* female had been found, as well as two females of *A. funestus* which showed at least 13 to 14 dilatations, indicating at least that number of ovipositions.

TABLE XI

Numbers of marked females of *Anopheles gambiae* recaptured at each calendar age in each physiological age group. Muheza, Tanzania, 1963–64 (after Gillies and Wilkes, 1965).

Calendar age in days (columns 1–34):

Physiological age	Abdominal stage	1	2	3	4	5	6	7	8	10	11	12	13	14	17	21	23	34	Mean calendar age
Pre-gravid		8	7																1·5
Nulliparous {	Fed		6	5	2	1													2·9
	Gravid			1	2														3·7
1-parous {	Fed				1	3	2	1											5·4
	Gravid						2	1	1										6·7
2-parous {	Fed							2	1	1									8
	Gravid								1			1							10
3-parous {	Fed										3		1						11·
	Gravid												1	1					13·
4-parous	Gravid												1						13
5-parous	Fed														1				17
6-parous	Fed															1	1		22
10-parous	Fed																	1	34

Despite the possibility noted above that extraneous conditions might prolong some cycles, the results of this study showed that the feeding behaviour of the population could be adequately described in terms of 3 to 5 days for the first cycle, and 5 days thereafter.

In the case of tse-tse flies, the problem has been tackled in much the same way by studies on *Glossina pallidipes* and others in Uganda (Harley, 1966, 1967a). Over a period of about eleven months just over 6 000 newly emerged females — less than 24 hours old — of three species of tse-tse were marked by Jackson's method (see page 96) and released. A total of eighty females were recaptured, and their physiological age categories and calendar ages recorded. The data with regard to thirty recaptured females of *G. pallidipes* are shown in Fig. 9, and indicate that each ovulation cycle occupies about eleven days.

As already noted (page 26) age grading techniques with tse-tse flies are less precise than with mosquitoes, and it is not yet possible to make further distinctions between females which have ovulated more than seven times, and whose age is eighty days or more.

With regard to male tse-tse flies which suck blood and play a role equal to that of females as vectors of disease, age grading methods based on physiological changes in the ovaries naturally do not apply. With male tse-tse, as well as females, the original method of age grading based on the extent of wing fray has recently been re-examined once more (Harley, 1967b), but it appears from this that it is still not possible to

relate wing fray accurately to calendar age, nor to determine the difference in mean calender age between the two sexes in wild-caught flies.

These two parallel studies on marking-release-recapture as a means
of illuminating the age composition of natural populations of anopheline
mosquitoes and tse-tse flies have both been rather handicapped by the
fact that conclusions have had to be based on the comparatively small
numbers of marked flies which were recaptured. In view of this, the fact

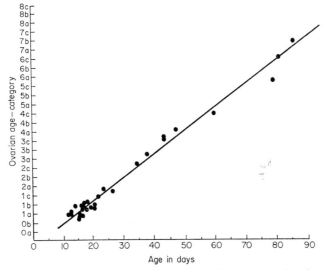

FIG. 9. Calendar age and ovarian age-categories of 30 recaptured females of *Glossina
pallidipes* (after Harley, 1966).

that so much new information has emerged from these investigations is
very encouraging, and strongly suggests that continued effort on those
lines would be rewarding. Such studies could perhaps help us to under-
stand much more about ecology of vectors in general, and particularly
the ecology of different age groups. In turn, this might be expected to
throw light on the extent to which different capture or sampling methods
are really securing a representative cross-section of all ages of the vector
population.

CHAPTER 9

SAMPLING OF FLEA POPULATIONS

The wingless insects of medical importance, such as fleas, lice and bugs, have a great deal in common with other arthropod vectors of disease such as ticks and mites, when considered from the point of view of sampling ecology. Associated with their common feature of being unable to fly, all these vectors tend to be very closely associated with their human and animal hosts and their immediate environment. When the host is man, the immediate environment may be the house or habitation in general, and the bed or sleeping quarters in particular. Some of the wingless insect vectors of disease are not entirely dependent on man as a host and may maintain themselves equally well on domestic or commensal rats or on wild rodents. In such cases the main aggregation sites are formed by the nests and burrows of the normal animal host. The distinction between winged and wingless vectors, useful from the point of view of the present review, is perhaps an oversimplification of the facts when applied to certain insects. For example, in the transmission of Chagas disease by Triatomid bugs, both the winged adults and the wingless nymphs act as vectors. Conversely, the common bed bug, Cimex, which is wingless in the adult stage and a widely distributed blood-sucking ectoparasite of man, is only an occasional or accidental vector of disease.

The problems posed by sampling such vector populations are very different in many ways from those encountered so far with the winged vectors. But nevertheless there are still many features in common, and it soon becomes evident that environment factors, and the behaviour of the insects themselves, play an equally prominent part in determining the choice of sampling method used by various investigators. Of particular interest in the present overall review is the fact that the varied sampling techniques for wingless vectors have been developed by specialists in the particular vector-borne diseases involved, and that consequently investigations have usually followed a course quite independent of and quite unrelated to similar studies on other vector-borne diseases.

Although it would be desirable to include all wingless insects as well as other arthropods such as ticks and mites in this appraisal, it is felt that the problems encountered in studying flea populations alone cover such a variety of conditions, and have been intensively studied for so long, that they are well suited to illustrate the ecological and behavioural background to sampling wingless vectors in general.

The earliest studies on flea populations were carried out in connection with the classic plague investigations in the early part of this century. Many of these investigations remain models of ecological and epidemiological study, and they established methods which are still in use today. Later, the study of flea populations extended to those species associated with wild or campestral rodents involved in the sylvatic plague complex. Additional information was provided by studies on rat fleas as vectors of murine typhus (Mohr, 1951). More recently, notable contributions to the problems of sampling flea populations in general have been made in the very different field of myxomatosis, where the European rabbit flea, *Spilopsyllus cuniculi* has received special attention.

FLEA COUNTS ON TRAPPED RATS

Ever since the association between rats, rat fleas and plague was first worked out, the problem of estimating the varying incidence and fluctuations in the flea population has been of primary importance in all subsequent studies.

The basic work of the Plague Commission (1906) in Bombay in the first decade of the century established the general finding that plague outbreaks, firstly in domestic rats and subsequently among humans, were usually associated with a high degree of flea infestation in rats. By trapping large numbers of domestic rats and collecting the fleas they carried, a figure was obtained for the average number of fleas per rat, that is the "flea index". At the time when the commission was working out the fundamental facts about plague transmission it was not yet realized that the *Xenopsylla* rat fleas of India belonged, not to a single species, but to three distinct species, *X. cheopis, X. astia* and *X. brasiliensis*. As these three species may play very different roles in plague transmission a specific flea index is now used, such as the *cheopis* index or the *astia* index. Of these the *cheopis* index, or average number of *X. cheopis* per domestic rodent — based on at least 100 rats and usually many more — is of the most universal interest, as *X. cheopis* still remains the most widely distributed and efficient vector of plague.

The conventional flea index makes no distinction between male and females fleas. Both sexes occur as blood-sucking ectoparasites on the rat host, and both are involved in the transmission of plague. Although in general both sexes of flea may be roughly equally represented in the majority of moderate or heavy infestations, the relative proportions of the two sexes do vary, and some of these variations appear to be related to environmental conditions. In the case of *X. cheopis* for example, field studies have shown that atmospheric temperature may be one of the factors concerned; Below 70°F females usually predominate, while over 75°F males are caught in greater number (Cole, 1945). The fact that populations of other fleas are liable to show very marked disparity in the proportions of males and females found on the host at certain seasons of the year, is well brought out in studies on the infestation of wild European rabbits with the rabbit flea, *Spilopsyllus cuniculi*, (Allan, 1956) (see page 118).

In some parts of the world it seemed that this *cheopis* index would prove, by itself, to be an accurate gauge of the risk of plague infection. For example in Ceylon, Hirst (1933, 1953) revealed a very close correlation between the plague incidence in Colombo over many years and the prevalence of *X. cheopis*, as follows. (See Table XII)

TABLE XII

Correlation between the plague incidence in Colombo over many years and the prevalence of *X. cheopis* (Hirst, 1933)

Cheopis index	Human plague Rate per 1000	Rodent plague % of total
Nil to 0·1	0·9	6·6
Over 0·1	5·8	29·9
Over 0·4	22·6	63·4

In parts of Ceylon other than Colombo about 90% of the bubonic and septicaemic plague cases detected occurred in districts showing a *cheopis* index exceeding 1·0. Before the appearance of plague in Colombo in 1914 rat fleas collected revealed *X. astia* alone, a relatively inefficient vector of plague. The first outbreak of plague revealed the presence of *X. cheopis*, and subsequent work indicated that under these conditions, when an *astia* area is invaded by *cheopis*, plague may become endemic when the *cheopis* index is near unity.

Outside Ceylon however, it appears that there are few places where the *cheopis* index alone is such a sensitive yardstick of plague endemicity. There are many places where there is no obvious relation between the changing incidence of plague and the varying degree of infestation of domestic rats with *X. cheopis*. In fact extremely high *cheopis* indices, perhaps 10 to 15 per rat, may persist in areas in which there has been no outbreak of plague for several years. In a very thorough plague enquiry in the Cumbum Valley in South India, (George and Webster, 1934) it was found that each year there is a peak of plague incidence in April and May although the *cheopis* index is in general lower at this time than the rest of the year. The decline in plague incidence sometimes sets in before the *cheopis* index begins to fall and the autumn rise in the *cheopis* index is not necessarily accompanied by an increased incidence in plague.

In the appraisal and assessment of the flea index it must be borne in mind that domestic rats, even in plague endemic areas, show a very wide range of infestation with *Xenopsylla* fleas, from zero up to a hundred or more. The low *cheopis* indices encountered in Colombo are such that many of the rats trapped had no fleas at all on them. Much higher indices may be encountered elsewhere; in a survey in Asyut in Egypt (Petrie and Todd, 1923) although not one plague infected rat was found in the course of two years, flea indices were very high, the average for a year from the three common rodents there, *Rattus norvegicus*, *Rattus rattus* and *Arvicanthis* being 10·0, 6·6, and 8·3 respectively. Even higher indices were recorded in rodents of feluccas, some months giving an average of up to 29 for *Rattus norvegicus*, *Xenopsylla cheopis* forming 90 to 100% of all domestic rodent fleas.

Individual rats may harbour unusually large numbers of fleas in general and *X. cheopis* in particular. In a rat-flea survey of Madras Presidency (King, *et al.* 1929) there were several records of rats — *Rattus rattus* — harbouring 29 to 30 *X. cheopis*, the maximum number of fleas of all species on any one rat in that survey being 70. Even higher numbers of fleas have been recorded by the Indian Plague Commission (1906) on rats dying of plague in Bombay. On one occasion 80 were recovered from a dying rat, and in another case about 300 were recorded from three dead rats. These latter figures might be regarded as exceptional as they were recorded at the height of a severe rat epizootic when the rodent population in Bombay had been decimated by plague, resulting in an unusual concentration of ectoparasites on the few remaining rats.

In interesting contrast to these earlier findings in India, more recent studies carried out in that country on essentially similar lines indicate that the *X. cheopis* index of domestic rats has reached a very low level and has remained there for several years. For example, in Calcutta, the *X. cheopis* index in 1951 was 0·03 to 0·32, in 1954 it was 0·6 to 0·58, and and in 1957, 0·01 to 0·17 (Seal, 1960). The lowest figure of 0·01 corresponds to the recovery of a single flea for every 100 rats examined.

The same investigation revealed the very significant fact that over the years there have been changes in the relative proportion of the common rat hosts of these fleas. In Bombay for example the proportion of *Rattus rattus* fell from 66% in 1910 to 23–24% in 1956. *Rattus norvegicus* also fell from 29% to 16% over the same period. The most striking change was in the proportion of *Bandicota bengalensis* which formed only 1% of the rodents in Bombay in 1910, but which rose to 49–50% in 1956.

The possibility that similar natural ecological changes are occurring elsewhere, and which may interfere with the host/parasite relationship or affect flea sampling in other ways, cannot be overlooked. In San Juan, Porto Rico, for example, there has been a decline in the *X. cheopis* rat flea population which has taken place even in the absence of D.D.T. dusting and in areas where rat control was unsuccessful (Fox, 1956); Fox and Garcia-Moll, 1961). At the same time there has been an increase in the other rat ectoparasites.

It has long been appreciated that rat fleas only spend part of their lives on their hosts, and that the number actually caught on trapped rats does not necessarily bear a direct relation to the flea population as a whole. But even if we were to assume for the sake of argument that all rat fleas spent all their lives attached to the host rat, the flea index would still be affected by so many variable factors as to render it very difficult to interpret in terms of the population as a whole. The effect of the most obvious variable, namely fluctuations in the rat population, has already been shown clearly in the Bombay figures quoted above. Evidence that these variables are now being more fully considered and recorded, is forthcoming from more recent flea-index studies in Pakistan (Rahman and Ahmad, 1963). Observations in Lahore are tabulated in the following distinct categories, viz. total number of rats collected; number of rats without fleas; rats infested, and average number of fleas per infested rat. The flea index for twelve months in 1961 expressed in a convenient way as

$$\frac{\text{Total number of fleas}}{\text{Total rats examined}} \quad \text{was} \quad \frac{675}{1355} \quad \text{or} \quad 0\cdot498.$$

Over the year, the mean number of rats infested was 20·7% showing that nearly 80% of the rats examined had no fleas at the time of examination.

If we really pursued this factor to its logical conclusion we would have to start by examining critically the methods used to sample the rat population before we attempt the much more complicated problem of the flea. Without digressing too much from the problem of the insect vector it is almost certain that the different methods of trapping employed do not yield uniform samples of the rat population (see page 115). An increase in the rat catch may be caused not only by an increase in the rat population but perhaps by a scarcity of food, making the baited traps more attractive, or by seasonal rainfall and floods driving the rat population nearer human habitations and trapping sites. For example in Bombay Province the conclusion of the plague off-season is marked by the advent of the rainy season which causes rats from outside shelters to herd into burrows indoors, greatly increasing the rat and rat flea populations in houses.

In interpreting the flea index or rather the *cheopis* index, the importance of taking the exact trapping site into consideration has been well brought out in a study of plague in Hawaii and the Hawaiian Islands (Eskey, 1934). In that area *X. cheopis* was present in fairly large numbers in every locality where plague was discovered, but it was found impossible to predict the susceptibility of the community to plague, the rapidity with which the disease would spread, or the severity of an epidemic from a simple calculation of the *cheopis* index. High *cheopis* indices were recorded in places that had remained immune to plague. The highest *X. cheopis* index recorded during the investigation was in a small village which had remained free of infection during the 12 years plague was in the vicinity. Here, over 1 000 *X. cheopis* were collected from twenty-five rats, all trapped in the same building. If trapping in this sector had been confined to rats caught outside the building the *cheopis* index would have been less than one, that is below a figure considered essential for the effective transmission of plague. Despite the misleadingly high infestation figures recorded for the rats inside the houses, the slight infestation of the rat population in general as judged by those caught outside the buildings did not permit the rapid spread of plague in this group of villages.

In using the flea index or the *cheopis* index as a guide to the degree of infestation of the local rodents one obvious difficulty is the occurrence of the occasional rat with an excessive number of fleas. In the example quoted above the small proportion of heavily infested rats caught inside buildings would have raised the *cheopis* index well above that representative of the rat population as a whole. In such cases the work in Hawaii showed that the "percentage of rats infested" had probably more significance than the usual form of flea index in indicating the extent of rodent infestation. Observations on three different species of rat flea showed a progressive rise in the average number of fleas found per rat with each increase in the percentage of infested rats (Fig. 10). The three species showed decided differences in the actual rate of increase of the indices, being least in the case of *Xenopsylla hawaiensis* and greatest for *Echidnophaga gallinacea*, but all of them show a remarkably progressive increase.

In the case of *X. cheopis*, an infestation of less than 30% was found to correspond to an index of less than 1, while an infestation from 35% to 50% indicated an index between 1 and 2. With an 80% infestation the high average of 6 *X. cheopis* per rat was reached. A similar relation has been reported elsewhere as for example in Porto Rico (Carrion, 1930) where the highest *cheopis* indices of 20 to 22 in April and May correspond to the highest percentage of rats infested, 90 to 100%. This relationship may possibly be of wider application.

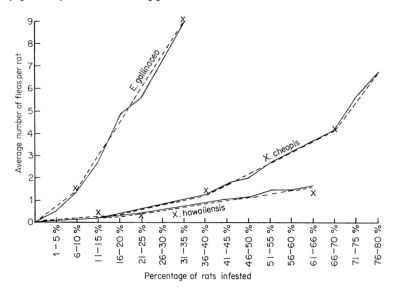

FIG. 10. Comparison of average number of fleas per rat, with percentage of rats infested in Hawaii (after Eskey, 1934): ——actual computation; - - - -Straight line between points marked X.

The relation between the percentage of rats infested, the *cheopis* index, and the situation in which rats were trapped is well brought out in Table XIII from the Hawaiian survey. Rats trapped inside buildings may of course include odd individuals that have come in from outside, but the rapid fall of flea infestation with increasing distance from the building suggests that this interchange of rodents does not take place to any great extent.

The need for distinguishing between the flea infestation of rats trapped inside buildings and those caught outside is further emphasized by the fact that the flea population on outdoor rats is much more affected by climatic changes. While the indoor flea population may continue to flourish, the number of fleas caught on outdoor rats may be greatly reduced by heavy rainfall, or by increased temperature. Presumably the shelter afforded by buildings in modifying the effects of high temperature is an operational factor. In Hawaii for example it was only at the higher altitudes, where the mean monthly temperature

TABLE XIII

Relation between percentage of rats infested, the *cheopis* index, and the trapping situation (Eskey, 1934).

Situation	% of rats infested	Index
Inside buildings	51%	3·57
Under, and within 50 ft	36%	1·66
51 to 200 ft	19%	0·50
201 to 500 ft	15%	0·36
Over 500 ft from buildings	7%	0·13

ranged from 67 to 60°F. and lower, that *X. cheopis* was encountered in appreciable number among field rodents.

The relationship of *X. cheopis* infestation to the trapping sites of the host rats does not necessarily apply to other species of rat flea. In fact in this particular survey another rat flea — *X. hawaiensis* closely related to *X. astia* of the Orient — showed a reversed relationship, infestation being greater on rats trapped over 1 000 ft from buildings than it was on those caught in close association with buildings. These differences and variation in the flea index in relation to buildings are probably greatly influenced by the nature of the main flea-breeding sites, which are principally indoors in the case of *X. cheopis*.

Any attempt to assess the true rat flea population by confining attention to fleas actually caught on rats will also be influenced by the species of rat. *Rattus norvegicus* usually yields more fleas and a higher *X. cheopis* index than *Rattus rattus*. In the same way, *Rattus norvegicus* in Colombo showed an *X. cheopis* index more than double that of *R. rattus* (Hirst, 1933). The fact that the relative proportions of the rat populations in towns may undergo dramatic changes, is well illustrated by the situation in Bombay (see page 105).

In a plague enquiry in Upper Egypt it was found that the two most abundant commensal rats were *Rattus rattus* and *Acomys niloticus* — the Cairo spiny mouse (Petrie and Todd, 1923). Although their domestic habits and habitats were very similar *Acomys* had only one-tenth the fleas carried by *Rattus*. This survey in Egypt also brought out another interesting factor bearing on the interpretation of the flea index. It was found that the infestation of fleas on rats increased with the size of the rat up to an optimum weight, and then decreased. Presumably the increase in infestation up to a certain point is related to the fact that the adult rat acquires more skill in ridding itself of its vermin. As young rats are usually caught more readily than old ones in cage traps, a further sampling error in estimating the flea population may be introduced (Eskey, 1930).

When commensal rats other than *Rattus* are taken into account, even greater differences may be found, producing wide differences in the composition of the flea population. While the emphasis has been on *X. cheopis* the plague flea par excellence, the many other species of flea liable to be found on these commensal rodents can no longer be dismissed lightly, as under certain conditions they may play an important part in plague epidemiology.

In many epidemiological studies on the transmission of insect-borne diseases, greater importance is usually attached to the population of infected or infective vectors than to the total or absolute population. However, with certain fleas, such as the common *Pulex irritans* which normally play no part in the transmission of plague from rat to rat or from rat to man, the concept of total population takes on a new significance in view of the fact that man-to-man transmission can take place in plague epidemics by means of mechanical transmission brought about by mass attack of such common fleas particularly associated with man.

SAMPLING OF FREE-LIVING FLEA POPULATIONS

So far we have been concerned with that part of the flea population which is actually found on trapped rats. Even assuming for the sake of argument that fleas spend the greater part of their lives attached to their hosts, the average number recorded at any one time in a sample of trapped rats would be greatly influenced by a multitude of variable factors which might on occasions give misleading ideas about the flea population as a whole. Apparent changes in the flea population according to season and climate could easily be caused by fluctuations in the numbers of rat hosts or changes in their activities.

When it is realized that the flea has an active life quite apart from its hosts, the problem becomes more intricate and at the same time more fascinating from the ecological point of view. Although the pioneer workers on plague were so concerned with rat flea indices, it has long been recognized that fleas have a life apart from the host, either in the rat nests and burrows, or as wandering fleas which have left the rat on its runways or at food stores. It was also early recognized that during the periods of plague epizootics when rat populations were decimated, large numbers of rat fleas left the bodies of dead rats and greatly increased the population of free-living fleas. In order to trap this wandering flea population, particularly in plague houses, where people were suffering from the disease, or where rats infected with plague had been found, the Indian Plague Commission in Bombay worked out simple methods which have a wide application. In one series of experiments de-flead guinea pigs were allowed to run about free in selected houses for 18 to 40 hours. Subsequent catches of up to 20 to 30 fleas on the guinea pig "sentinels" were frequently recorded, and on one occasion 106 fleas were found on two of these guinea pigs. At that time the existence of three species of *Xenopsylla* in India was not recognized, and no exact records of flea identification are available for this experiment. Presumably most of the fleas trapped were *Xenopsylla*, and possibly the majority of these were *X. cheopis*. In another series of experiments bait animals were put in cages in these houses and surrounded by a six-inch tanglefoot to trap approaching fleas.

The use of de-flead guinea pigs was further developed in the Cumbum Valley investigations in South India (George and Webster, 1934). Bait animals in pairs were allowed to run overnight in houses; in addition individual burrows were tested by introducing a de-flead rat tethered by wire and left overnight. This method has also been used to detect the presence of wandering rat fleas in imported food stuffs, for example by allowing guinea pigs to run loose in bullock carts of grain (George and Webster, 1934). The examination of floor sweepings in houses or huts in the course of plague investigations has also proved a profitable method of picking up *X. brasiliensis* the dominant flea on domestic rodents in many of the more elevated parts of Africa (Davis, 1953, 1964).

There is no doubt that under certain conditions this free-living population of rat fleas can reach enormous proportions. In the Egyptian plague work referred to above, search in plague houses abandoned by the owners revealed a huge residue of *X. cheopis*. Guinea pigs liberated in the rooms picked up large numbers of fleas and when the walls were dismembered and broken up it was clearly shown by the prodigious number of fleas released that the nests and burrows were the main centres of flea concentration, and the main reservoir of house infestation. Normally these reservoirs are in inaccessible sites in the foundations of the thick mud walls of the local houses (Petrie and Todd, 1923).

The ability of free-living fleas to find and attach themselves to the occasional host has been very well demonstrated experimentally in the case of the rabbit flea (*Spilopsyllus cuniculi*). Groups of fleas were marked in distinct ways by removal of the terminal subsegment of the tarsus, and were scattered in an enclosure into which flea-free rabbits were subsequently released. Two hundred and seventy-four fleas were scattered in a 2000-sq. yd enclosed pasture, and 3 flea-free rabbits were released one hour afterwards. After 18 hours the rabbits were recaptured by netting, combed several times to remove all fleas, then replaced and the procedure repeated. By the end of 8 days the overall recovery of liberated fleas was 45% (Mead-Briggs, 1964c).

It has now been established that in some parts of the tropics a considerable free-living population of *X. cheopis* may exist at a time when this
flea can rarely be found on local rats. This was well shown in Madagascar
where a human plague-infected case carried the disease to an isolated
area where, despite the fact that there had been no plague and no rat
epizootics, he was able to pass the infection to other members of the
family. In a collection of 126 fleas from the plague-infected house, 122
were *Pulex irritans* and only 4 *X. cheopis*. However, by using an improved
technique involving sifting and flotation of dust and debris (Estrade,
1934) large numbers of free-living *cheopis* were recovered from grain
stores in the house, and also from rice shops, the collection yielding 1313
cheopis and 158 *Pulex irritans* (Girard, 1943).

The pioneer work of Hirst in Ceylon had stressed the probable
importance of *X. cheopis* in imported grain as a means of introducing
plague infection, but the extent of this free-living population had not
really been tested thoroughly.

Under certain conditions fleas other than *Xenopsylla* can play an
important part in plague transmission, particularly in transmitting infection from man to man. The fact that some species of flea, while
individually poor vectors, can still bring about mechanical transmission
by mass attack, has given a new significance to these populations of
wandering fleas closely associated with man but rare or absent from
domestic rats. Dakar, in West Africa, is an endemic focus of plague,
and suffered from an epidemic in 1944. While *X. cheopis* is the dominant
species on the rat there, another flea — *Synosternus pallidus* — is abundant
in local grass and wooden huts, on floors and on human bodies. Sheets
of fly-paper were used to sample the flea population in these huts, and
succeeded in trapping anything up to 270 fleas per hut. The composition
of the flea population on rats and in huts is shown in Table XIV.

TABLE XIV

Composition of flea populations on rats and in huts (Kartman, 1946).

	Rats	Huts
X. cheopis	85·7%	0·7%
E. gallinacea	5·4	8·4
S. pallidus	3·5	90·7%
S. canis	5·4	0·2

During the plague season the *X. cheopis* population as judged by the
infestation on rats, remained low with a *cheopis* index of about 0·96, and
it was concluded that *S. pallidus* in view of its abundance and close

association with man was probably the main man-to-man vector of plague (Kartman, 1946). A somewhat similar role is undoubtedly played by the common house flea, *Pulex irritans*, under certain conditions. It may be the chief agent in a few places like the highlands of Ecuador (Eskey, 1930) where human plague persists in the absence of *X. cheopis* and where there is a considerable amount of person-to-person transmission of plague among the mountain Indians. In such a place *Pulex* is a pest everywhere in houses, blankets and clothing, and may even be found on rats in some localities.

SAMPLING OF SYLVATIC RODENT FLEAS

While we now know that the free-living and wandering phases of rodent fleas may form a significant portion of the whole flea population, as far as domestic rodents are concerned it has not been possible to form anything more than a vague idea of the relative proportion of these phases. We would like to know much more about the distribution of the flea population between the rat and the nests and burrows. Those who have worked with commensal rats and domestic plague have been perhaps unduly concerned with fleas actually found on the rat. In sylvatic plague studies on the other hand it has long been appreciated that many of the fleas of wild rodents tend to infest nests and burrows to a greater extent.

Studies have been made in parts of California, heavily populated with Ground Squirrels, *Citellus*, one of the main rodent reservoirs of sylvatic plague in the Western United States. Fleas were collected from the burrow openings by means of cotton wool, as well as from the rodents themselves. Of 7500 fleas collected, only about a dozen were species other than *Diamanus montanus* and *Hoplopsyllus anomalus*. With these two species there was a close correlation between fluctuation in the burrow population and the flea index of the rats (Fig. 11) and it was concluded that collection of samples from the burrow mouths would give just as accurate an idea of flea populations as flea indices (Stewart and Evans, 1941).

This still leaves rather uncertain the extent of the nest population which other investigators (Eskey and Haas, 1940) indicate may be quite considerable. For example, of 708 *Citellus beechyi* examined, the average number of fleas per animal was 26·3. In 39 nests the average number of fleas per nest was 69, and it was shown that there was a direct relation between the number of fleas infesting rodent hosts and the numbers infesting the nests.

A more comprehensive study of the distribution of different fractions of the flea population has been carried out on another wild rodent species, *Citellus pygmaeus*, involved in the sylvatic plague complex in

Northern Europe (Mironov *et al.* 1963). Counts were made of the flea
— principally *Neopsylla setosa* — on the rodent occupant of the burrow,
at the entrance of the burrow and on the runs, and in the breeding
chamber. Over the whole period of investigation 90·8% of the fleas were
found in the breeding chamber, 6·4% on the rodent occupant, and 1·4%
each at entrance and on runs.

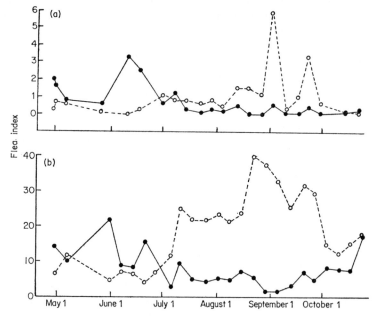

FIG. 11. Fluctuations in populations of fleas: ● *Diamanus montanus*, and ○
Hoplopsyllus anomalus, (*a*) taken at burrow mouths and (*b*) on the host ground squirrel
(*Citellus*) (after Stewart and Evans, 1941).

In one of the first detailed studies on the plague problem in the South
African veld (Mitchell *et al.* 1927) it was realized that while the "flea
index" may be of some value when dealing with commensal rodents and
their fleas, it was a very different matter to formulate an index for fleas,
such as *Dinopsyllus lypusus* and *Chiastopsylla rossi*, whose hosts are gerbilles
and other sylvan rodents, and which live mainly in the nests, resorting
to the hosts only for the purpose of feeding. Gerbilles make use of
several nests, migrating at frequent intervals from one to another. As
many of the abandoned nests contain numerous fleas, this factor also
has to be taken into account in forming some estimate of the total
population. A somewhat modified "flea index" appropriate to these
conditions was worked out taking into account the following data: (*a*)
Total fleas taken from rodents, from occupied nests and from un-

occupied nests; (*b*) Number of rodents captured — allowing an additional two animals as equivalent to each abandoned nest. On the basis of this data, the index was found to vary from approximately 1 to 3 in different localities. This work also brought out the fact that the distribution of fleas in the nests and burrows may be determined by the nature of the nest; some species of rodents build substantial nests in which fleas congregate, while others make nests which do not retain fleas or their larvae to the same extent (see also Kleim, 1963).

The limitations of the conventional flea index under such conditions have been recognized by other workers. Machiavello (1950) for example suggested the use of an "absolute flea index", i.e. "the average number of fleas, including nest fleas, that theoretically can feed on one rat" in the area surveyed. In order to obtain this wider range of data, it is necessary to know the total number of nests and rodents, and the relative number of fleas per rodent and per nest. An attempt to obtain quantitative data of this kind has been carried out in Hawaii (Haas, 1966), in which subterranean nest boxes were used to study the nesting habits of the rodents concerned in cane fields. Boxes were arranged in a grid system, all boxes being inspected at monthly intervals, when rodents were trapped and de-flead, and adult fleas removed from the nests. After examination, the nests with rats and their fleas were returned to the nest boxes. In addition, trapping was carried out on four successive nights a month, and fleas counted on the rats. Theoretically, it was considered that knowing the total number of nests per hectare, and the mean number of rodents per nest, plus the mean number of fleas per trapped rodent, it should be possible to calculate the total number of fleas in nests and on rodents per hectare. In practice considerable possibilities of error became apparent due to various uncontrollable environmental factors.

As already mentioned, one of the difficulties in basing estimates of flea populations on collections made on the rodent host is that the trapping or sampling of the host itself may involve a whole new range of variables. The results which are obtained with rodents which are live-trapped may differ from those obtained when the rat is killed, whether by snap-trap (Miles *et al.* 1957) or by Cyanogas (Fox and Garcia Moll, 1961). In the latter case there may be loss of fleas between the death of the rat, and its actual collection for flea examination (Deguisti and Hartley, 1965). The actual method of de-fleaing rats may also involve a further source of error if not carried out with care and attention (Balthazard and Eftekhari, 1957).

With regard to the question of making accurate estimates of the host rodent population as an essential first step towards realizing the ideal of an absolute flea index, the methods used in studying the common

North American rabbit flea (*Cediopsylla simplex*) on the American Cotton Tail rabbit (*Sylvilagus*) are illuminating (Mohr and Lord, 1960). The rabbit population was estimated in two areas by trapping, marking and release. This was supplemented by estimates based on the proportion marked in the hunter's bag, an estimate not affected by trap-shyness on the part of the rabbit. Other supplementary indices were: rabbits harvested; rabbits per gun hour, and rabbits per 100 trap-nights. The attempt to determine the absolute rabbit population in each area was combined with studies on the infestation rate of rabbits with fleas, and counts of the fleas on each infested rabbit. Although it was not possible to extend the final results beyond a comparison of the two areas, the investigation points the way to a new and more critical quantitative approach to an old ecological problem.

Perhaps the experience gained in other rabbit flea investigations noted above (Mead-Briggs, 1964c) may suggest that this complex sampling problem might be illuminated by experiments involving the release of known numbers of marked fleas in such plots, or by the rather more ambitious release of known numbers of marked rabbits or rodent hosts.

To return to the commensal rodents, there is every indication that the nest population of fleas may on occasions be enormously greater than might be supposed from the flea indices, and the relationship is neither as direct or as regular as with sylvatic rodents. In addition, the distribution of fleas between host and nest in typical nocturnal rodents like *Rattus rattus* might well be expected to differ from that of say *Arvicanthis*, a semi-domestic rodent which plays an important part in plague transmission in Kenya, and which is essentially diurnal in its habits (Heisch *et al.* 1953).

Numerous attempts have been made in the laboratory to investigate this question of the partition of flea populations between host, and nest or bedding. Results have by no means been consistent, and may well remain difficult to interpret until such time as the question is investigated more systematically in the field. There seems to be wider scope for the methods which were developed in plague studies in South Africa (Davis, 1939, 1964) in which whole warrens of the rodents — in this particular case the Gerbille, *Tatera* sp—are excavated foot by foot to determine the distribution and numbers of fleas throughout the system. The value of this technique has recently been emphasized in sylvatic plague studies in India and in Kurdistan in which step-by-step excavation of rodent burrows and galleries was carried out, in the course of which continuous flea collections were made on the debris, as well as on the fur of both live and dead rodents (Balthazard and Bahmanyar, 1960—Klein, 1963).

The experimental approach to flea populations and flea movements in the field has been developed elsewhere in quite a different context, namely in studying the behaviour of adult bird fleas (Bates, 1962). By constructing artificial burrows, at the mouths of which traps — containing water and detergent — were installed, the migratory movements of fleas by day or by night could be checked. In this same investigation, an "artificial bird" in the form of a polythene bottle containing water at 50°C. and covered with the skin of a bird, was used for collecting fleas from debris on the ground by moving it over the ground and stroking the soil in order to disturb and attract the fleas.

As described above, it is often very difficult in the case of fleas of commensal rats to decide whether seasonal fluctuations in the numbers of fleas per rat represent true fluctuations in the flea population, or whether they are determined by seasonal or other changes in the population of the rat hosts. Studies on the more stable populations of field rodents involved in sylvatic plague indicate that numbers of fleas really do undergo regular seasonal fluctuations. For example, in the ground squirrel of the Western United States (*Citellus beechyi*) — one of the most important rodents in the sylvatic plague cycle — the two dominant fleas show quite different seasonal trends (Eskey and Haas, 1940) to such an extent that surveys restricted to one season of the year can give misleading ideas of the relative abundance of the fleas concerned (see Table XV).

TABLE XV

Seasonal fluctuation of number of fleas (*Diamanus montanus* and *Hoplopsyllus anomalus* on *Citellus beechyi* (Eskey and Haas, 1940).

Period	*Diamanus montanus* index	*Hoplopsyllus anomalus* index
December–February	27·9	3·6
March–May	28·8	1·6
June–August	11·9	9·5
September–November	11·7	10·5

In the Great Salt Lake desert, seasonal fluctuations of *Hoplopsyllus anomalus* have also been noted (Parker, 1958) this species being most abundant on ground squirrels during the warm summer months, while none were collected from December through to March. The most abundant flea species reached a maximum in December.

A somewhat similar alternation has been recorded from South Africa with the fleas of the Gerbille (*Tatera brantsi*) (Mitchell *et al.* 1927; Davis, 1939) an important rodent of the sylvatic plague reservoir. During the summer *Xenopsylla enidos* is most abundant on the gerbille, but during the winter it is replaced to a great extent by *Chiastopsylla rossi* which is almost absent in summer.

RELEVANT STUDIES ON SAMPLING AND ECOLOGY OF RABBIT FLEAS

The introduction of myxomatosis into Britain in 1953 stimulated considerable interest in the European Rabbit Flea (*Spilopsyllus cuniculi*) the most likely principal vector of the disease (Lockley, 1954; Allan, 1956; Muirhead-Thomson, 1956b; Rothschild, 1960; Rothschild and Ford, 1965b). Subsequently, this interest lead to a more penetrating study of the biology of the rabbit flea and of its peculiar physiological relationship with its rabbit host (Mead-Briggs and Ruage 1960) (Mead-Briggs, 1964a; Rothschild and Ford, 1965a). The most significant disclosure of the latter investigations was the finding that although the female rabbit flea may infest rabbits of both sexes and all ages, it is only on the pregnant female rabbit that the female flea can undergo maturation of the ovaries, leading eventually to egg-laying in the nest of the doe. The sexual cycle of the female rabbit flea is controlled by the reproductive hormones of its rabbit host.

It would be interesting to speculate about the likely repercussions of this physiological work on the host/parasite relationship of the rabbit flea in so far as it might apply to other fleas and other hosts, and to see what bearing these newly disclosed factors might have on the ecology of flea sampling. However, for the present purpose it will be sufficient to select from all this material one or two observations of particular relevance to quantitative aspects of sampling flea populations in general, in addition to those examples which have been noted earlier in this chapter (page 111).

The first investigation is of special interest in that the greater part of it was carried out on a natural wild-rabbit population in Scotland which had not yet been affected by the advent of myxomatosis (Allan, 1956). In this investigation rabbits were trapped each fortnight from late December onwards, and samples of 10 full grown males and 10 full grown females were examined for fleas. In this way information was obtained not only with regard to seasonal changes in the flea infestation on rabbits, but also with regard to seasonal changes in the relative incidence of fleas on male and on female rabbits.

The results presented in Fig. 12 show a gradual increase in the number of fleas per rabbit between fortnight two and four, i.e. between 8 January and 5 February, followed by a large increase at fortnight six in early March. Initially, this marked increase occurred mainly on the doe, but by fortnight ten — the end of April — the population was equally divided between the rabbit sexes. This is followed by a sudden drop in the flea population, which remains relatively low on the rabbit until there is another increase towards the end of the year, with a repetition of the same marked peak in the following spring.

In this investigation a separate record was kept of the distribution of fleas on different parts of the host, viz. the ears, the head and the body. *Spilopsyllus* is usually regarded as having a marked preference for the rabbit's ear as an attachment site, and long before myxomatosis stimulated interest in England, people who were in the habit of trapping or

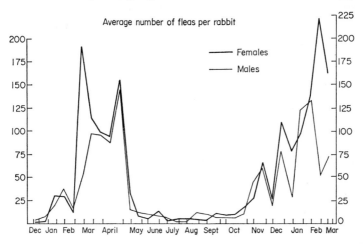

Fig. 12. Seasonal abundance of rabbit fleas (*Spilopsyllus cuniculi*) on wild rabbits in Scotland, and distribution of flea population between male and female rabbits (after Allan, 1956).

shooting wild rabbits were familiar with the clusters of these fleas commonly seen inside the tip of the ear. At certain times of the year however, fleas can be found on head and body, although the ear in general remains the main locus of infestation.

When this distribution was examined critically and quantitatively, some very striking features were revealed (Fig. 13). While the fleas on the ear represent nearly 80% of the total infestation for all fortnights, a very marked change in proportions occurred in January when the body population increased to a peak accounting for over 71% of the total. The subsequent fall in body population was accompanied by an increase in the head population to a peak corresponding to nearly 50% of the total. By the beginning of March, both body and head populations had fallen to a very low level, nearly all fleas now being found on the ears only. The body fleas remained at a very low level throughout the summer, until the sharp increase was repeated the following winter. The head population however, tended to persist throughout the summer, with irregular peaks.

Of additional interest from the point of view of vector behaviour in relation to sampling is the observation that the fleas found on the ears normally react in a very different way from those found on other parts of the host body. Fleas found on the ear, the main site of attachment in wild rabbits, are usually firmly attached and difficult to disturb or dislodge. In contrast those on the head are restless and easily disturbed. Even gentle blowing on the top of the rabbit's head will induce fleas to come to the surface of the fur, some of them being sufficiently excited as to jump right off the host (Muirhead-Thomson, 1956).

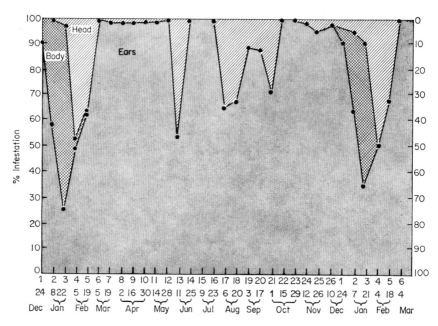

FIG. 13. Seasonal distribution of rabbit fleas (*Spilopsyllus cuniculi*) on ears, head and body of rabbit host (after Allan, 1956).

When taken in conjunction with the seasonal changes in distribution pattern of rabbit fleas on the host's body, it seems quite likely that this associated difference in flea behaviour could play an important part in determining the extent to which interchange of fleas can take place between different hosts, and also the extent to which fleas are liable to leave the body of the live host and become free-living for a time in nests or burrows. In this respect it seems significant that the low rate of flea infestation on wild rabbits during the summer months (Fig. 12) should follow a period when there is a peak incidence of the more active and restless fraction of the flea population on parts of the rabbit other than the ears.

A further example of investigations on rabbit flea populations is

provided by a straightforward study on infestation rates of rabbits —
mainly myxomatous — with rabbit fleas (Mead-Briggs, 1964b). In a
survey of 279 rabbits from every county of Great Britain, it was found
that the infestation rates for individual rabbits varied from 0–273, and
that 4% of the rabbits were flea-free. The details of these findings are
shown in Table XVI. It is interesting to note that although very high
infestation rates were recorded in some individual rabbits, over 60% of

TABLE XVI

Frequency distribution of fleas, *Spilopsyllus cuniculi*, on a wild
population of the European Rabbit, *Oryctolagus* (after
Mead-Briggs, 1964).

Number of fleas on rabbit	Number of rabbits	Number of rabbits as % of total
0	9	3·8
1–10	109	45·8
11–20	38	16·0
21–30	22	9·2
31–40	16	6·7
41–50	14	5·9
51–60	5	2·1
61–70	8	3·4
71–80	3	1·3
81–90	1	0·4
91–100	2	0·8
101–110	2	0·8
111–120	5	2·1
121–130	0	0·0
131–140	1	0·4
141–150	1	0·4
150+	2	0·8

the animals examined had 20 fleas or less on them. This same survey
produced useful information about the relationship between flea infestation rates and size (weight) of the host rabbit. The results Table XVII,

show that in general the larger rabbits record larger numbers of fleas (see also page 109).

(see also page 109).

TABLE XVII

Mean infestation rate of *Spilopsyllus cuniculi* on wild rabbits in relation to unpaunched weight of host (after Mead-Briggs, 1964).

Weight range in grams	Number of rabbits	Number of fleas	Mean number fleas/rabbit	Number of rabbits without fleas
Under 750	14	130	9·3	0
751–1 000	29	251	8·7	3
1 001–1 250	70	1 620	23·1	1
1 251–1 500	78	2 165	27·8	2
1 501+	47	1 485	31·6	2
	238	5 651	23·7	9

In contrast to the wide range of moderate to heavy infestations recorded in these different weight groups, rabbit fleas are rarely taken on very young rabbits, particularly those removed from rabbits' nests, despite the fact that these nests are usually swarming with fleas (Muirhead-Thomson, 1956). Of 18 rabbits ranging between 3 and 10 days old taken from nests, not a single one was found with fleas. Laboratory observations showed that hungry fleas will feed readily on the youngest rabbits found — 3 to 4 days old, and quite hairless — so presumably the absence of fleas from the body of the baby rabbit does not rule out the probability that in nature fleas feed on them at some time or other, but do not remain attached for more than a brief period.

The several different studies on the European rabbit flea have undoubtedly revealed important new factors which must be taken into account in studying populations of fleas in general, particularly the vectors of urban and sylvatic plague. The ecological aspects of sampling such flea populations are very complex, and there is still a long way to go before anything more than a tentative interpretation of crude sampling data can be attempted in terms of absolute flea populations.

In this connection it is worth noting that the rabbit flea infestation data discussed above — together with data regarding flea infestation of rats in Burma and of small rodents in England — has been the subject of a rather different type of analysis whose object is to find a mathematical pattern to the distribution of such ectoparasites among hosts (Williams, 1964). This analysis starts with the general observation that when the

average number of parasites is low, the number of hosts with one parasite is greater than the number with two: the number with two parasites
is greater than the number with three and so on. The attempt is then
made to find a mathematical model to fit the observed data in order
ultimately to throw light on the mechanism of population balance. In
this particular analysis it was evidently difficult to draw conclusions
regarding the data from Burma, but in the case of fleas on small rodents
in England, and rabbit fleas in Scotland, the data was found in general
to conform closely to a logarithmic series.

While it would obviously be instructive to discuss this mathematic
approach in more detail, the subject is rather outside the scope and
competence of the present review. What is more in keeping however, with
the nature of this review on sampling and ecology, is the fact that the
author of the work referred to above, draws attention to the inherent
error in collecting parasites from hosts, particularly at low infestations
where the loss or overlooking of a single flea may lead to an excess of
zero counts.

CHAPTER 10

SUMMARY AND DISCUSSION

The main object of this review has been to attempt an overall assessment of fundamental problems in the ecology of insect vectors of disease, with particular reference to the capture or sampling methods on which so much of our existing knowledge of ecology is based. In this way it is hoped to extract from the many different branches of vector entomology, the particular experiences and conclusions most likely to be of value in formulating a more co-ordinated approach to common problems in the future. It might therefore be profitable to analyse several different aspects of the same general problem in turn, and see what emerges. The bulk of this general discussion will be concerned with the dipterous vectors of disease, as it is felt that any extension to the ecology of wingless vectors could not be made on the basis of flea studies alone, but could only be justified if consideration was given to many other wingless vectors, such as lice, ticks and mites, which have not been dealt with in this present review.

BEHAVIOUR, ECOLOGY AND SAMPLING

One of the most significant points emerging from the present review is that population sampling, ecology, and vector behaviour are intimately bound up with each other. One of the primary objectives of sampling adult vector populations is to establish a clearer understanding of vector ecology, and to investigate behaviour patterns, especially those relating to the insects' role in the transmission of human or animal disease. But it is quite clear from the many independent and unrelated lines of investigation which have been pursued in different fields of medical entomology, that the initial selection of sampling methods to be adopted has in many cases been determined by the behaviour and ecology of the insect concerned.

In general, the behaviour pattern of each vector species has tended to determine the extent to which the population can most conveniently be sampled, whether in its active biting phase, in its inactive resting phase, or in flight, and has also determined the extent to which these different fractions could be most suitably sampled during the day, at dusk, by night, or at some other specified time. The behaviour of the adult vector with regard to its selection of particular concentration or aggregation sites has also been a prime factor in deciding the choice of sampling method by different investigators, and in determining the validity of the same sampling technique under different environmental conditions or at different seasons of the year. Any modification in behaviour pattern which manifests itself in a greater or lesser degree of aggregation, in increased or decreased attraction to man or other hosts, or in increased or decreased flight activity, will obviously influence sampling data based on those activities.

Environmental factors such as climate, vegetation, availability of host animals and so on, may exercise a profound influence — whether acting singly or in combination — on vector density and distribution, and may also have a marked influence on vector behaviour. From this it follows that when sampling methods based on what appears to be a uniform behaviour pattern in one environment, are applied to a different environment they may fail to provide a reliable yardstick of differences which exist, or changes which are produced, in the vector population as a whole. In the same way, the continued use of the same sampling method in a particular area through periods when climatic and other factors undergo marked change, may give a very distorted estimate of changes in vector density during that period. In extreme cases these continued sampling records may even cease to have any relative value, and become nothing more than a series of disjointed observations.

It seems from this that the first step towards progress would be a much keener appreciation of the extent to which current knowledge about vector ecology, distribution and behaviour has been based on data provided by capture or sampling methods which themselves have evolved on lines determined by the behaviour and ecology of the insect concerned. In order to obtain adequate numbers of adult vectors for comparative purposes, it is very natural that full use has been made of aggregation sites where populations are concentrated at select feeding or resting places. However, the investigator cannot really assume that these are the only aggregation sites, as there may well be situations in space or in time where hitherto undetected concentrations of adult insects occur, and where population sampling might well lead to the need for completely revising ideas about ecology and behaviour. In addition, limitations of time have too often made it impractical to sample vector populations in areas or habitats where they are sparsely distributed, but in which nevertheless a very high proportion of the total vector populations might exist.

An evaluation of the possible sources of error in the fundamental methods employed in ecological studies is also closely bound up with the exact type of information which the investigator hopes to obtain.

DEFINITION OF OBJECTIVES IN VECTOR SAMPLING

In establishing present knowledge about the ecology of insect vectors of disease, one of the striking features is the great range and variety of methods used to capture insects and to sample adult populations. This is evidently by no means peculiar to the field of vector entomology, but is also characteristic of a wide range of general insect investigations recently reviewed and assessed by Southwood (1966).

Many factors, including personal interests or individualism on the part of different investigators working in comparative isolation, have contributed to this variety of ecological methods. But perhaps the major contributing factor has been the wide range in the nature and objectives of the different ecological studies.

Some sampling methods have been designed mainly for survey to determine the geographical distribution of vectors, their seasonal abundance and local distribution, or their association with different types of vegetation (tse-tse) or with different types of breeding ground (mosquitoes, *Simulium*). Other sampling methods have been aimed primarily at collecting sufficient material for dissection for parasites, or for virus recovery. Others have been mainly developed in connection with behaviour studies or with establishing the degree of contact between vector and man, while still others have arisen from the need to evaluate vector control campaigns. Some confusion has been caused by developing sampling methods for one purpose, and then using these methods indiscriminately in order to deal with quite different requirements.

From the material discussed in this review it seems that it would be useful for entomologists to define clearly the objective of each sampling method used; to say whether they are concerned with the vector population as a whole, or only with a particular fraction associated with a certain habitat or with a certain host. It would be important to be clear in one's mind whether the use of the bait catch is designed to illuminate host preferences, or whether it is intended to indicate changes in vector population density. In the same way it is important to know whether the main objective of a particular sampling method is to study the composition of the population with regard to its age grouping and infection rate, or as an estimate of the total population. On such a basis there would appear to be an urgent need for the pooling of the vast and varied experience of many different entomologists, so that the entomologist in one discipline could readily find out how his particular sampling problem is dealt with in some other discipline. Instead of gaining information only of a general nature, he could thus hope for something more specific, such as the sampling method used by other entomologists to investigate presumed vector eradication, or to study sampling methods which others have used to detect changes of host preference under insecticide pressure.

SAMPLING BASED ON HUMAN OR ANIMAL BAIT

One of the most widely used sampling techniques in several different fields of insect vector study is the routine or standardized capture of insects attracted to exposed bait — human or animal. The use of the human "fly-boy" or bait/collector is practiced with all groups of biting flies, mosquitoes, tse-tse flies, black-flies, sand-flies, tabanids, etc. It is noteworthy that this particular method is one very liable to be influenced by behavioural features of the vector, as well as by environmental factors. The number and composition of the bait catch are liable to wide variations, often unpredictable, according to the nature and number of the bait animals, what periods of the day the bait is exposed; the relation between this period and the diel biting cycle of the insect; the duration of exposure; whether the bait is moving or stationary; the relative attraction of different units of the bait; (e.g. some humans are much more attractive to biting insects than others); the presence of competing sources of animal attraction in the neighbourhood; and the density and activity of the vector. In addition there are numerous environmental factors liable to influence the bait catch, such as wind intensity and direction; air temperature and humidity; rainfall; the proximity or otherwise to breeding grounds of the vector concerned; and last but not least, the skill of the insect collector himself.

Captures on bait carried out indiscriminately, without due regard to all these factors, may still yield a "good" catch of insects, but the interpretation of the data obtained in terms of vector ecology or abundance may too easily become a matter of sheer speculation or personal opinion.

However, provided the objectives are clearly defined to illuminate particular aspects of vector ecology, captures based on live bait can be extended or standardized in such a way as to minimize at least some of the variables mentioned above. One of the most striking trends in this direction has been the evolution of the 24-hour catch in the classical studies on mosquitoes in relation to jungle yellow fever and other insect-borne viruses in Central Africa (page 52). By means of a standardized method of collecting on standardized number of bait units at regular intervals throughout the day and night, many of the variables due to hour-to-hour or day-to-day variations in vector activity and abundance can be counteracted. In addition, the arrangement of bait platforms at different heights above the ground makes full allowance for differences in vector activity or flight at all levels from ground to forest canopy. Where this type of 24-hour catch has been carried out systematically, and with due regard to adequate supervision, it has lead to marked progress in understanding vector ecology, not only with regard to pre-dominantly night-biting mosquitoes, but also with many day-time biters such as tse-tse flies and Tabanids.

Progress in the development of this round-the-clock sampling technique has been mainly due to the energy and sustained efforts of a small number of dedicated investigators. It would be encouraging to think that captures on live bait — including humans — could be perfected on automatic or mechanical lines in such a way as to obviate the human element involved in the actual capture of the insect biting or attempting to bite the bait. It is difficult to visualize how this could be done without some sort of trap or trapping device, and this innovation itself would almost certainly introduce new variables and obstacles.

Progress in the improvements of techniques based on capture of vectors on exposed bait have been parallelled by developments in the use and design of animal-baited traps. Where man himself provides the attractant bait, perhaps the major developments in mosquito studies at least have been along two rather distinct lines, viz. the use of the man-operated bed net trap, or baited net trap, and the evolution of the experimental hut/window trap technique. The latter technique, with its advantages in the way of automatically trapping mosquitoes leaving occupied huts through the windows or eaves throughout the night from dusk to dawn, has provided a wealth of new information on particular aspects of mosquito ecology. However, the extension of this sampling method to study vector populations in ordinary human-occupied houses — as distinct from specially constructed experimental huts — has encountered many operational difficulties, mainly due to the extreme diversity of pattern of human activities within and around their own dwellings.

With regard to developments in the use and design of traps baited with animals other than man, it is clear that the sampling is exposed not only to all the variables noted above in connection with routine collecting on exposed bait, but also to a whole new range of factors connected with the structure, efficiency and operation of the trap itself. Developments in the design of traps baited with small animals has been particularly marked in the case of the smaller biting flies, black-flies, midges, and sand-flies. A great deal of this progress has taken place along individual or independent lines, unconnected with similar efforts elsewhere. With those groups of insects, which have many problems in common, it seems that much would be gained through greater co-ordination of effort, and fuller attention given to something which is so rarely done objectively, viz. a full and fair comparison of different types of trap (The "Smith" trap, the "Jones" trap, the "Brown" trap, etc.) in one and the same place. If this were done in a few representative areas or countries by a team of investigators, including those who have special experience of each specialized technique, the outcome might well be the development of a much more efficient sampling method, incorporating the best features of the different designs submitted.

Many of the basic principles involved in trap construction and efficiency have been closely examined in studies on blow-flies and house-flies. Although the attractant bait for those non-blood sucking insects usually takes the form of carrion or some attractive food mixture, such questions as siting of the traps, "trap idiosyncrasy", and ingress and egress by the insects concerned, are all equally applicable to the design and operation of live-baited traps for biting insects.

The need for more co-ordination of effort is again very evident in the case of sampling based on visual traps, of the "silhouette" or "artificial-animal" type. Considerable progress has been made along several independent lines, and it is clear that some of the careful investigations made on non-vector biting insects in temperate climates have escaped the notice of medical entomologists in the tropical field. Although these various traps are designed to sample primarily the biting or attacking portion of the population, as far as that fraction is concerned the techniques do eliminate many of the sampling variables attending the use of live bait, including man, and for this reason alone, should merit further studies with different vectors.

EXTENSION OF THE SAMPLING SPECTRUM IN ECOLOGICAL STUDIES

One of the most interesting trends with regard to tse-tse flies and anopheline mosquitoes is the increasingly critical approach to catching or sampling methods, and the growing tendency to test different types of sampling method synchronized so as to be able to compare in a more objective manner the advantages and disadvantages of each. The complete reliance long placed by the malaria control entomologists on the "house catch" is parallel to the long preoccupation of tse-tse workers — in some areas — with the "fly round". The studies in both these groups of vectors now appear to be extending in the same spirit of critical inquiry. It is interesting to speculate how much this is due to the fact that the possibility of control or eradication of malaria and trypanosomiasis, or their vectors, has necessitated new critical standards for evaluating the reactions of these vectors to insecticides or other control measures.

In striking contrast to this is the fact that, despite consistent work over many years on the forest-dwelling culicine vectors of virus in Central America and Central Africa, it is only comparatively recently that there has been any extension or amplification of sampling methods corresponding to those developed with tse-tse and anopheline mosquitoes.

As the feasibility of controlling these forest culicines is too remote for present contemplation, it is difficult to avoid the conclusion that the possibility of applying effective vector control appears to be a very necessary stimulus towards the development of new ideas and more critical standards.

With many groups of winged biting vectors it appears that although a variety of sampling methods are in use, there are still large and significant fractions of the vector population which are not adequately represented. In general it is the resting inactive fraction of the population existing in outdoor haunts which has not yet been fully represented by available methods, especially that phase between blood-meals, when the blood is digesting and the ovaries of the females developing.

A great deal of progress in this direction has been made in studies on some anopheline mosquitoes and tse-tse flies, but much more remains to be done even within those groups. In the case of forest culicine mosquitoes, and with black-flies, we are still remarkably ignorant about the nature of those outdoor resting sites, and still very far from developing methods for sampling what in many cases must be a considerable proportion of the total existing vector population. Lack of knowledge on this aspect of ecology even extends to one of the best-known insect vectors of human disease, namely *Aedes aegypti* the classical "Yellow fever mosquito" of the Americas. In contrast to the wealth of information about this species under laboratory conditions, its outdoor resting habits still remain practically unknown (Schoof, 1967).

Despite the recognized difficulties in the way of sampling these outdoor populations in many tropical environments, it is difficult to avoid the conclusion that in most cases lack of success can most simply be explained by the fact that the problem has not been tackled with sufficient vigour, or on a sufficiently large scale. A particularly encouraging example of what can be achieved by a vigorous application of relatively simple methods was provided some years ago by the studies on outdoor resting populations of mosquitoes in Colombia (page 36). By means of a portable cage placed over unit area of vegetation, and a repellent chemical to disturb the resting mosquitoes, large number of samples could be obtained representing a wide range of environmental conditions. A somewhat similar approach has been the use of tent traps to sample blow-flies resting on vegetation outdoors at night, in which the flies are trapped automatically when they leave their night time resting places at dawn and are attracted to the light coming through the transparent apex of the tent.

There seems no obvious reason why these methods should not be capable of further development in such a way as to sample not only larger areas of several hundred square metres at a time, but also be capable of dealing with a wide range of vegetation types, including thickets or even small trees.

A good example of the way in which problems in vector ecology are now being tackled on a very much larger scale than hitherto attempted is provided by investigations connected with the possible application of the sterile male technique to tse-tse control in Rhodesia. (Dame *et al.* 1965). In order to build up a high population of tse-tse introduced into a natural habitat, an acre (4047 sq. m) of woodland was encircled by a 17 ft wall of cotton gauze cloth attached to a wire framework. The hope that this enormous roofless cage would prevent introduced tse-tse from emigrating was not fully realized, but at least the technique illustrates the need to tackle big problems in a big way.

The need to sample this outdoor population takes on an added significance when it is borne in mind that with many vectors this may well be the sole fraction of the adult population which survives unfavourable seasons of the year — particularly combinations of prolonged drought and high temperature as is the case with *Phlebotomus* in the Sudan. It is this elusive fraction which may be responsible for the well-known explosive population outbreaks which so often mark the return of favourable conditions, and which so often follow periods in which the vector seems to have virtually disappeared.

Recent trends in the field of tse-tse and mosquito studies have been characterized, not only by an increasing use of different sampling methods based on different principles, but also by the critical comparison of these methods in the same place and at the same time. This is an encouraging trend which can no longer be ignored in other fields of vector study, in particular the study of *Simulium* vectors of onchocerciasis and the sylvan culicine mosquitoes which are vectors of various virus diseases. The almost complete reliance long placed on the bait catch in these latter two fields now appears to be well out of step with recent advances in the field of tse-tse and anopheline studies.

Of particular value have been comparisons of sampling methods based on quite different physiological phases in the life of the vector, such as the active hungry or biting phase with the inactive phase digesting the blood meal between feeds, as has been done with tse-tse flies and anopheline mosquitoes. Studies on the culicine mosquito vector of urban filariasis in Rangoon and Colombo have also been noteworthy in illuminating the ecology of a hitherto almost neglected fraction of the population, namely the gravid females attracted to ovipositing sites. Similarly, comparison between the active biting phase and the population in flight — as sampled by means of non-attractant net traps, suction traps or sticky traps — has provided further insight into the ecology of sampling. Perhaps rather less instructive are comparisons involving methods in which the nature of the attractant is ill defined. A good example of this is the Morris "animal trap" which may perhaps be visually attractive to hungry tse-tse because it resembles an animal in appearance, or which on the other hand may be attractive to those tse-tse which are simply seeking a dark resting site.

In order to extract the maximum information from such comparisons of sampling methods, it is clear that a great deal of careful planning and design is required, and that ideally the different methods should be compared over a wide range of vector density.

In view of the bias inherent in most sampling methods based on various forms of attractant, whether live bait, light traps, carrion or carbon dioxide, increasing attention is being given to devising non-attractant methods for sampling vector populations. The main object of such methods is to obtain random unbiased samples of the popula-tion in flight by means of trapping methods employing static nets, as in the Malaise trap, (Townes, 1962; Breeland and Pickard, 1965), sweep nets attached to moving vehicles; suction traps and sticky traps. The development of such techniques would appear to offer the best prospects for minimizing or eliminating bias in sampling populations of vectors; at the same time it is difficult to avoid the conclusion that the results to date are in general disappointing. This may perhaps be due to the fact that the methods have not been tested on a sufficiently large scale to allow for the fact that even in flight, many tropical insect vectors of disease occur at low densities, or in a high state of dispersion. In addition, the problem of bias cannot be overlooked even in these non-attractant methods, the bias in this case being the human one whereby the operator, consciously or unconsciously, determines the place, time and other conditions under which these sampling methods are operated. It is extremely unlikely that the vector population in flight exists for any length of time in a uniformly distributed pattern, and accordingly even a non-attractant sampling method may have to allow for questions of aggregation and concentration.

At the present stage of progress it would seem that non-attractant sampling methods can best be developed and evaluated when used in conjunction with a variety of other methods in a broad spectrum ap-proach to sampling. The value of this approach has already been demon-strated by work on mosquitoes in Colombia (page 36) and North America (page 66) in which sampling by means of mechanical sweep nets attached to vehicles has been compared with more conventional methods based on attractants such as light or live bait.

SAMPLING OF THE SEXES

A great deal of our present knowledge about the ecology of adult mosquitoes is based on capture or sampling directed at the female portion of the population. Neglect of the male mosquito population — apart from swarming and mating activities — has undoubtedly been conditioned by the fact that the males do not suck blood or transmit disease and therefore "don't count". The author himself, despite a long association with mosquito ecology, is aware of having possessed this peculiar blind spot with regard to anopheline mosquitoes. This neglect is also evident in the majority of investigations on culicine mosquitoes, where by far the most prevalent sampling method used in ecological studies has been based on attraction to live bait, a technique in which males only play a very minor or accidental role.

An interesting contrast to this is provided by ecological studies on savannah tse-tse flies. Although both sexes of tse-tse suck blood and transmit disease, for many years one of the most widely used methods for sampling and studying these insects dealt almost entirely with the male population attracted to human bait in the "fly-round". Current methods for sampling tse-tse populations now cover a wide range, involving both sexes, but in the case of practically all other groups of insect vectors such as mosquitoes, black-flies, sand-flies and midges, basic sampling methods take little account of the non-blood sucking male population. In keeping with this, present knowledge about the ecology of males, as distinct from females, of these different vector groups, is still very inadequate.

The need to deal with this very obvious gap in knowledge has taken on a new urgency in view of the fact that the successful application of new methods of genetic control based on the male sterilization principle require more accurate information than is available regarding the density, dispersal, mating activity and longevity of the male population.

With regard to the more general problem of sampling aimed at estimating absolute population density, it seems that further studies on male ecology might well lead to the development of sampling techniques more precise and reliable than those currently based on the female fraction of the population. The life of these non-blood sucking males is not complicated by the physiological processes associated with blood feeding, ovarian development and ovipositing, which play such a dominant role in the ecology and behaviour of the female. It seems possible therefore that — apart from activities concerned with swarming or mating — the male population might eventually be shown to exist in a much more uniformly dispersed pattern than the widely fluctuating and fluid female fraction.

Increasing attention has recently been given to the flower and nectar feeding behaviour of both male and female mosquitoes and black-flies, (Wenk, 1965; Lewis and Domoney, 1966), and there is evidence of specific preference based on chemical attraction. Further investigations in this field might provide something which would be of great value in extending the sampling spectrum, namely an automatic trap for male mosquitoes in which the chemical attractant was sufficiently strong to produce a high degree of aggregation in areas where the population would normally be greatly dispersed and difficult to detect.

PHYSIOLOGICAL CYCLES AND AGE-GRADING

The present approach to problems of vector ecology and the epidemiology of vector-borne disease is being increasingly influenced by the development of more precise techniques for estimating the physiological age of individual insects. This progress has been most marked in the case of mosquitoes, both anopheline and culicine, and tse-tse flies, but considerable advances have also been made in studies on black-flies (*Simulium*). A great deal of this work has been concerned with two main problems, firstly, the use of age-grading data to provide estimates of the daily mortality of vector populations in nature — particularly populations under pressure of control, and secondly, to study the relationship between age of the vector and its degree of infection with different development stages of parasites.

However, there is another aspect of this work which is of more immediate concern in connection with the interrelated problems of vector ecology and vector sampling, that is to see if the new age-grading methods reveal any tendency for the behaviour pattern of vector species to change with age.

The very extensive work carried out on the ecology of anopheline mosquitoes in recent years has shown clearly that the physiological changes which take place in the female during the 3 to 4 days of the gonotrophic cycle, have a profound influence on its behaviour, and consequently on its availability by different sampling methods. The movements and resting habits of the female mosquito may show marked changes during the digestion of the blood meal and the development of the ovaries, and these changes may become evident well before the final phase of complete ovarian development leading to oviposition. This pattern tends to vary from species to species, with the consequence that one particular phase in the cycle may be over-represented by one sampling method, and under-represented by another. The availability of these different phases in the cycle may be further affected by changes in the duration of the cycle, such as those induced by seasonal change or by climatic differences.

In the case of insect vectors such as tse-tse flies in which males as well as females are blood suckers, a somewhat similar cycle exists in the case of the male. Male tse-tse pass through a hunger cycle which starts with a full blood meal and continues through stages where blood digestion is accompanied by accumulation of fat. This stage is then followed by depletion of fat reserves and eventually by "hunger". As described on page 25 these different physiological stages of male tse-tse differ in their availability according to the different sampling methods used.

In view of these established facts regarding behaviour changes within each cycle, it would obviously be important to establish whether or not corresponding changes in pattern are produced as a result of increasing age and the cumulative effect of repeated ovarian cycles (or "hunger" cycles in the case of male tse-tse). The application of the simple age grading technique which subdivides the female sample into two categories — nulliparous and parous — has shown that certain sampling methods record a much higher proportion of nulliparous females than other methods, thus confirming that the young and recently emerged females of the mosquito population differ in behaviour and availability from the older population in general. What would be even more useful to know is whether there are differences in behaviour within the parous group itself, to the extent possibly that the older multi-parous females which have undergone several ovarian cycles, might develop significant differences in feeding and resting habits from the younger parous females which have completed only one or two cycles.

While it would be reasonable to believe that the extending application of the advanced age-grouping methods would throw direct light on this problem, nothing very significant one way or the other has emerged so far. However, as this advanced technique is still a very difficult and highly specialized one, making great demands on time, perhaps it is rather early to expect rapid progress towards this goal.

Further studies on this aspect of age grading should undoubtedly be stimulated by the fact that the existence of such changes in behaviour pattern have already been established in the case of tse-tse flies. Older females of *Glossina morsitans* have been shown to be less attracted to man as bait than younger females in the first three weeks of life. The additional information that older female tse-tse are less susceptible to insecticide than young ones gives added significance to this observation, and suggests that the special ecology of the older or ageing vector would well repay further study from the point of view of the epidemiology and control of vector-borne disease in general.

SAMPLING AT HIGH AND LOW DENSITIES

The concept of high vector density or low vector density is a very vague and arbitrary one. The tse-tse workers, especially those concerned with *G. palpalis*, principal vector of human trypanosomiasis due to

T. gambiense, have long been acutely aware that many of their vectors can still transmit disease at such low densities as to be barely detectable by their refined methods. By these standards the present concepts of vector scarcity or "absence" which commonly exist in the field of anopheline studies, are in most cases extremely crude and uncritical.

Ideas about vector abundance and vector scarcity are usually directly influenced by the efficiency of the sampling methods used. Methods devised at a time when vectors are abundant, and readily taken in great numbers at all times, are the very ones which are often liable to reveal serious limitations at low densities. There is already abundant evidence in the field of anopheline ecology at least to show that mosquitoes resting in the undergrowth in a state of dispersion on the border line of detection — perhaps not more than one per hundred square metres — can still amount to an impressive total population in terms of hectares or square miles. How much this scattered population can be concentrated on an isolated attractant such as human or animal bait is difficult to say, but there is some evidence to show that only quite a limited area may be tapped in this way.

The same uncertainty applies to the effective radius of the various kinds of traps constructed to resemble animals, or to provide shaded resting places, or both. Although the general thinking of tse-tse workers in this respect is sufficiently advanced to provide a stimulating example to workers in other vector fields, it appears that future needs in all vector control or vector disease control programmes will demand an entirely new approach to the study of vector fluctuations and composition at levels which are still epidemiologically significant, but are at the same time not readily detectable by the methods at present available. What lines this investigation will take is difficult to say at the moment, but perhaps the use of some mechanical method of trapping based on chemical attractant will prove more sensitive and more reliable than the various man-operated methods which are so much more likely to be subjective and exposed to the human factor.

One of the greatest obstacles to the interpretation of sampling data in terms of the true population is that it is never possible in nature to refer these figures to a known — as distinct from an estimated — population. Perhaps this could only be done satisfactorily by a large cage experiment in which a known number of vectors is liberated or allowed to emerge. The artificiality of such cage experiments, no matter how large the cage, has had a discouraging effect on such lines of investigation; but exact and controllable experiments of this kind have been done with house-flies and may yet prove a necessary adjunct to the development of more critical field methods for other insects. With the present increased emphasis on the quantitative approach to insect and animal ecology, it is possible that improved methods of measuring output of adult vectors from well-defined breeding habitats, may give a more accurate idea of production per unit area in unit time, and thus provide a rather more substantial backdrop to sampling data which would otherwise remain empirical and speculative.

The question of the detection of vectors at what might be called "sub-diagnostic levels" is closely bound up with the question of actual vector eradication, and the reappearance of vectors in areas from which they have been reported absent for long periods. These are vital questions of concern to all those who are engaged in the study or in the control of vector-borne diseases. A much closer integration of different disciplines and a much keener appreciation of common objectives may play a decisive part in determining how rapidly entomologists can progress in this direction.

SAMPLING PROBLEMS IN THE INTERPRETATION OF BLOOD—MEAL DATA

The identification of the source of blood meal in engorged insect vectors, by means of the precipitin tests, has long been recognized as one of the most accurate and valuable techniques available for studying the host preferences of biting insects in nature (Weitz, 1956). For many years it has been used extensively in numerous investigations on the ecology of the mosquito vectors of malaria, and in more recent years its use has been extended to tse-tse flies in Africa (Weitz and Glasgow, 1956; Glasgow et al. 1958; Jordan et al. 1960, 1961; Weitz, 1963) and to the culicine mosquito vectors of various types of viral encephalitis in Colorado and California (Reeves et al. 1963; Tempelis et al. 1965, 1967). The technique has also been applied in a more limited way to identifying the blood meals of engorged black-flies (*Simulium*) collected at light traps in Scotland (Davies et al. 1962).

For the more precise identification of blood from closely related animals or groups of animals, a more discriminative test (inhibition of agglutination) has been developed and has proved particularly valuable, for example, in distinguishing man blood from monkey blood in the course of investigations on forest mosquitoes involved in the transmission of simian malaria and human malaria.

The widespread application of the precipitin test has provided a great deal of accurate information about the identity of the mammals and birds on which insect vectors actually feed in nature, in addition to man. At the same time it has become increasingly evident that the interpretation of all this data is by no means simple and straightforward, and that in comparatively few cases has a clear-cut and consistent feeding pattern been revealed. Many of the insect vectors which regularly feed on man also feed on domestic or other animals to a variable extent depending on the availability of these alternative hosts. With regard to those vectors which also feed to a considerable extent on wild animals such as mammals and reptiles (tse-tse flies) or birds (culicine mosquito vectors of encephalitis), it appears that the feeding habits may fluctuate enormously according to rapid changes in the density or proportion of their wild hosts.

Apart from these unpredictable, and usually uncontrollable natural events, is the fact that all information on host preference as determined by the precipitin test is based on samples of insect vectors captured by a variety of methods and under a variety of conditions. If the blood-meal data is based on a sample of the vector population collected in a single type of habitat, e.g. a human dwelling, or in a stable or animal shelter, precipitin testing will usually show quite clearly that the feeding preference is heavily biased in favour of the dominant host in the collecting site. At the same time, tests based on samples of engorged vectors collected well away from the influence of human habitations and human settlements, may reveal a very low proportion with human blood, and may well be biased in favour of some known or unknown wild host in the vicinity of the collecting site. When engorged vectors can be found in a variety of resting sites, both indoors and outdoors, the question then arises as to which sample most closely represents the feeding pattern of the vector population as a whole, and what weight can be attached to each sample of different origin in trying to arrive at a consolidated estimate of host preference.

These difficulties in interpretation are particularly well illustrated in a recent appraisal of one of the most comprehensive series of precipitin tests yet attempted, namely, the co-ordinated work carried out in the ten years, 1955 to 1964, on the host preferences of anopheline mosquitoes in connection with the world-wide malaria eradication programme (Bruce-Chwatt *et al.* 1966). This appraisal is based on over 124 000 tests — of which 94% gave positive results — covering 92 different species or species-complexes of *Anopheles*. The main aim of this work was to provide a clearer idea of the basic blood preferences of all the major and secondary vectors of malaria in order to obtain a more accurate idea of their vectorial capacity. In addition, the plan of testing was designed to indicate any change of feeding pattern brought about by the pressure of insecticide applied on a vast country-wide scale, and also to clarify the status of the different members of species-complexes regarding their feeding preference for man and their role as vectors of malaria.

In reviewing the wealth of information provided by this unique survey, the authors conclude that in the majority of cases the results give a valid indication of the proportion of bites taken on man at the time and place of sampling, but they also point out that before expressing these results in the form of a more precise "human blood index", certain reservations must be considered due to difficulties in sampling. Two of the most notable sampling variables — which are in fact likely to be encountered in a wide range of insect vectors in general — were considered to be (a) lack of knowledge of the true biotopic distribution of the blood-fed females, and (b) the low efficiency of available sampling techniques in outdoor shelters. In addition were two equally important factors more closely associated with the actual vector control programme, viz. (c) mosquito scarcity (natural or due to spraying) rendering some monthly samples too small for comparative analysis, and (d) non-inclusion of blood meals from mosquitoes which, in sprayed areas, are killed after feeding but before the hour of collection.

The authors' final recommendation that further field research is required to overcome the sampling difficulty refers specifically to the mosquito vectors of malaria, but is clearly one which is also valid for allied problems in the much wider field of vector ecology in general.

A combination of precipitin test and agglutination-inhibition test has been widely used in tse-tse fly studies in order to clarify the feeding pattern of the different species with regard to animal hosts other than man. In many ways it might appear that the sampling problems involved are rather less complex than with the domestic and semi-domestic mosquito vectors of malaria discussed above, in so far as tse-tse flies feed and rest almost entirely outdoors. Provided the sampling of this outdoor population of engorged flies is carried out in a comprehensive manner, and with due regard to seasonal representation, there appears to be a good chance that the blood-meal survey may give, for each vector species, a reasonably accurate idea of the host preference of the population as a whole. There are one or two unusual features of tse-tse ecology however, which need closer examination in this respect. First of all, both sexes of tse-tse are blood-suckers, and the feeding habits of both males and females have to be considered. In earlier blood-meal surveys in East Africa, sampling was confined to the partly-fed males attracted to a moving party of 2 to 3 men, no engorged females being taken in this way. Sampling of the active males was later supplemented by samples of gorged tse-tse of both sexes taken resting on vegetation, and it was concluded that results based on male tse-tse alone were in fact applicable to the population as a whole.

Latterly, increasing attention has been given to sampling confined to the resting population of engorged flies taken in vegetation and forest, and this work has revealed that both sexes are not always equally available even in the quiescent resting phase. While *Glossina palpalis* for example produced samples in which the two sexes were taken in approximately equal numbers, with both *G. longipalpis* and *G. pallicera*, more engorged males than females were collected. The suggestion made to explain this disparity was that the partially gorged males are more attracted by the presence of the catchers, and therefore more liable to be caught than the females.

The interpretation of precipitin test data regarding tse-tse is particularly subject to the variables mentioned above concerning day-to-day fluctuations in the density and composition of the wild animal populations on which these flies feed to a large extent. Nevertheless, these refined techniques have provided a much more definite picture not only of the wide range of animals on which tse-tse feed in nature, but they have also confirmed earlier crude observations that certain dominant animals such as zebra appear to be completely unattractive to tse-tse as a source of blood. Perhaps further discriminative application of precipitin testing may in time help to confirm and extend observations already referred to regarding the change in feeding pattern — of female tse-tse at least — according to the age of the vector.

IMPACT OF MAN-MADE ENVIRONMENTAL CHANGES

Ideas and conclusions about the ecology of insects vectors of disease need constant review in the light of man-made environmental changes, particularly those which are characteristic of so many developing tropical countries. Such changes imposed on the environment may have a pronounced effect on the validity of the capture or sampling techniques which provide the basic tools on ecological work. A striking example of the way in which such a drastic alteration may interfere with long-practised methods of sampling adult vectors is provided by the introduction of house-spraying with D.D.T. for the control of the anopheline mosquito vectors of malaria. Many of the anophelines concerned use human and domestic animal shelters as day-time resting places, and consequently the day-time catch of resting mosquitoes was a long-established method of assessing vector density. When the inner walls and other surfaces of such habitations are sprayed with D.D.T., the day-time house catch falls to zero or near zero, and remains at a very low level for weeks or even months after a single treatment. In the early, and less critical, phase of D.D.T. application in malaria control, this apparent dramatic drop in house catch was interpreted as direct evidence that the D.D.T. treatment was achieving almost complete elimination of the vector population. However, it is now recognized that with several anopheline species, a varying proportion of the population attempting to rest on the inner treated surfaces of the house, are irritated by the D.D.T. residues and may escape from the house un-harmed, without having absorbed a lethal dose of insecticide (Muirhead-Thomson, 1960b; de Zulueta, 1964). In its extreme form this behavioural effect of the D.D.T. application is to achieve a shift of resting population from indoors, where the population is aggregated and easily detected, to outdoors where the population becomes widely dispersed and difficult to detect in vegetation and other natural shelters. With some species this disrupting effect is further accentuated by a reduction in the actual number of mosquitoes entering treated houses, perhaps due to some irritant or repellent effect of the D.D.T. deposits at windows, eaves, or other entry points. In this case man-imposed environmental changes have rendered an apparently reliable sampling method quite unreliable, and in some instances, grossly misleading.

It seems quite likely that other man-made environmental changes, perhaps less drastic or less obvious, will also interfere with the validity of long-accepted sampling methods. In many developing countries in the tropics the old, dark, mud and thatch type of rural habitation is giving way gradually to brick walls and corrugated iron roofs, giving lighter and airier structures less suitable as day-time shelters for mosquitoes and other vectors associated with man and his habitation. In the same way, the rapid urbanization and suburbanization in those countries is changing the environment, and the relative proportion of man and his domestic animals, in such a way that sampling data based on conventional captures at vector aggregation sites — resting places or feeding sites — are liable to become mere figures with no real meaning in terms of vector population as a whole.

Capture or sampling data form the main foundation for so much of present knowledge of vector ecology and the epidemiology of vector-borne disease, that a keen appreciation of all the factors involved in sampling must be regarded as an essential step towards a clearer under-standing of all the complex problems involved.

QUANTITATIVE ECOLOGY AND THE LIFE TABLE CONCEPT

In November 1966, a small international group of entomologists and ecologists met in Geneva to discuss the subject of Mosquito Ecology. This group was not limited to mosquito specialists, but included authorities in the much wider field of animal ecology and population studies. In addition, unlike many of the larger conferences and seminars on insect ecology, this representative group had the responsibility of pre-paring a joint report and recommendations, the final draft of which had to be approved by all participants before the meeting finally broke up. This unusually significant report (W.H.O. 1967b), crystallizes not only the very diverse range of experience brought to bear on this particular problem by the actual members, but also embodies a wide coverage of up-to-date information put at the disposal of the meeting in the form of unpublished working documents specially prepared for this purpose by collaborators in different countries.

It is a rather unusual event for entomologists and ecologists to meet under such conditions, in which joint decisions and recommendations have to be made, which are acceptable to the group as a whole. As a consequence, although the meeting was immediately concerned with the subject of mosquito ecology, many of the final issues are equally relevant to allied problems in vector ecology as a whole. In view of this it would be profitable to summarize the more general findings of this scientific group, and then see what bearing these have on the material reviewed in this present book.

The group first of all drew attention to the fact that "understanding of the epidemiology of the diseases involved and assessment of control programmes depend on accurate measurements of population, and knowledge of the causes of the growth and decline of mosquito populations is essential for the integration of different approaches to vector control". They pointed out that much of the available information on mosquito ecology was of a qualitative or relative nature, and that consequently "there was still a serious lack of quantitative information on many aspects of mosquito ecology".

The group considered that better methods for the measurement of numbers of mosquitoes would provide a better understanding of population dynamics, and that the full potentialities of applying mathematical or statistical analytical methods for studying and forecasting vector populations and patterns of disease frequency would remain unrealized without more accurate measures of population size and understanding of population dynamics.

The report then draws attention to the fact that various ecologists, particularly those concerned with forest entomology, have developed methods for the construction and analysis of a life-table, or life-budget, and that this new approach should be applied to mosquitoes. Life-tables describe, in terms of absolute population size or density, the number of individuals in a generation or cohort passing through each stage, and the contribution of these individuals to the next generation. In these connections the report emphasizes that "the central requirement of life-table work is absolute population estimates", and that wherever possible such estimation should be made by more than one method in order to ensure that part of the population is not totally overlooked. The report concludes that as much of the work on mosquitoes has emphasized relative methods, studies aimed at the construction of a life-table will initially have to be based largely on techniques developed for other animals.

It is evident that the opinions and resolutions of this report concerning the points summarized above have a direct bearing on ecological problems common to all insect vectors of disease in general. One of the purposes of this present book has been to draw attention to various environmental and behavioural factors which have an influence on the capture or sampling methods used for adult insect vectors of disease. It seems that in keeping with the increasing emphasis on the quantitative approach to ecology, the first step in progress would be the full awareness of the variable factors likely to affect the validity of basic data obtained by current sampling techniques. From this the next step would be to measure or minimize these variables, either by adequate replication of sampling under different conditions, by widening the sampling spectrum in order to embrace other aspects of vector ecology, or by devising methods — perhaps involving much more refined marking-release-recapture techniques — aimed at expressing sampling data in terms of absolute population density.

From material presented in this present book there seems little doubt that the initial stimulus could be provided by a closer co-ordination of effort, and pooling of knowledge and resources, among the many investigators in the diverse fields of vector ecology. A further incentive, in keeping with the opinion of the W.H.O. scientific group, would be a keener awareness of advances in progress on fundamentally similar problems of population studies in other fields of entomology or animal ecology.

In order to obtain a more composite picture of vector ecology, and a more complete contribution to the life-table, a similar precise quantitative approach would be necessary with regard to the immature stages of vectors, viz. eggs, larvae, and pupae. This particular subject is outside the scope of the present review, but is clearly one due for critical reappraisal.

Many field entomologists, acutely conscious of practical difficulties and obstacles in the way of planning and executing valid field experiments under difficult conditions in developing countries, may feel that the standards of accuracy implicit in the life-table concept are idealistic and unattainable. However, despite the known and acknowledged difficulties in obtaining more exact quantitative data on vector ecology under those exacting conditions, there is little hope of progress without having a main aim or objective in mind.

If measurements and observations in the field can be geared to this ultimate requirement of establishing accurate life-tables and all that they signify, it will give an entirely new meaning and purpose to all future investigations. This new outlook will almost certainly lead in turn to much higher critical standards, not only in development and operation of sampling methods, but also in the interpretation of the essential sampling data on which so much of our knowledge of vector ecology depends.

BIBLIOGRAPHY

Allan, R. M. (1956). *Proc. R. ent. Soc. Land.* A**31**, 145–152. A study of the population of the rabbit flea *Spilopsyllus cuniculi* (Dale) on the wild rabbit, *Oryctolagus cuniculus*, in north-east Scotland.

Anderson, J. R., and de Foliart, G. R. (1961). *Ann. ent. Soc. Am.* **54**, 716–729. Feeding behaviour and host preferences of some black-flies (*Diptera: Simuliidae*) in Wisconsin.

Anderson, J. R., and Poorbaugh, J. H. (1964a). *W.H.O./Vector Control/102.64.* p. 4 World Health Organization, Geneva. Refinements for collecting and processing sticky fly-tapes used for sampling populations of synanthropic flies.

Anderson, J. R., and Poorbaugh, J. H. (1964b). *J. med. Ent.* **2**, 131–147. Observations on the ethology and ecology of various Diptera associated with Northern California poultry ranches.

Andrewartha, H. G., and Birch, L. C. (1960). *Ann. Rev. Ent.* **5**, 219–242. Some recent contributions to the study of the distribution and abundance of insects.

Andrewartha, H. G. (1961). *Introduction to the study of animal populations*, 281 pp. Methuen and Co. Ltd.

Arevad, K. (1965). *Entomologia expt. appl.* **8**, 175–188. On the orientation of house-flies to various substances.

Ashcroft, M. T. (1959). *Trop. Dis. Bull.* **56**, 1073–1093. A critical review of the epidemiology of human trypanosomiasis in Africa.

Balthazard, M., and Eftekhari, M. (1957). *Bull. W.H.O.*, **16**, 436–440. Techniques de récolte, de manipulation et d'elevage des puces de ronguers.

Balthazard, M., and Bahmanyar, M. (1960). *Bull. W.H.O.* **23**, 169–215. Recherches sur la Peste en Inde.

Barr, A. R., Smith, T. A., and Boreham, M. M. (1960). *J. econ. Ent.* **53**, 876–880. Light intensity and the attraction of mosquitoes to light traps.

Barr, A. R., Smith, T. A., Boreham, M. M., and White, K. E. (1963). *J. econ. Ent.* **56**, 123–127. Evaluation of some factors affecting the efficiency of light traps in collecting mosquitoes.

Bates, J. K. (1962). *Parasit.* **52**, 113–132. Field studies on the behaviour of bird fleas. Behaviour of the adults of 3 species of bird flea in the field.

Beadle, L. D. (1959). *Am. J. trop. Med. Hyg.* **8**, 134–140. Field observations on the biting habits of *Culex tarsalis* at Mitchell, Nebraska, and Logan, Utah.

Beesley, W. N., and Crewe, W. (1963). *Ann. trop. Med. Parasit.* **57**, 191–203. The bionomics of *Chrysops silacea* Austen, 1907. II. The biting rhythm and dispersal in rain forest.

Bennett, G. F. (1960). *Can. J. Zool.* **38**, 377–389. On some ornithophilic blood-sucking Diptera in Algonquin Park, Ontario, Canada.

Bennett, G. F. (1963). *Can. J. Zool.* **41**, 831–840. Use of P^{32} in the study of a population of *Simulium rugglesi* (*Diptera: Simuliidae*) in Algonquin Park, Ontario.

Bertram, D. S., and McGregor, I. A. (1956). *Bull. ent. Res.* **47**, 669–681. Catches in the Gambia, West Africa, of *Anopheles gambiae* Giles and *A. gambiae var melas* Theo. in entrance traps of a baited portable wooden hut with special reference to the effect of wind direction.

Bertram, D. S., and Samarawickrema, W. A. (1958). *Nature, Lond.* **182**, 444. Age determination for individual Mansonioides mosquitoes.

Bidlingmayer, W. L. (1961). *Ann. ent. Soc. Am.* **54**, 149–156. Field activity studies on adult *Culicoides fureni.*

Blanton, F. S., Galindo, P., and Peyton, E. L. (1955). *Mosquito News.* **15**, 90–93. Report of a 3-year light trap survey for biting Diptera in Panama.

Boorman, J. P. T. (1960a). *W. Afr. med. J.* **9**, 111–122. Studies on the biting habits of the mosquito *Aedes (Stegomyia) aegypti.* Linn. in a West African village.

Boorman J. P. T. (1960b). *W. Afr. med. J.* **9**, 235–246. Studies on the biting habits of 6 species of culicine mosquito in a West African village.

Bracken, G. K., Hanec, W., and Thorsteinson, A. J. (1962). *Can. J. Zool.* **40**, 685–695. The orientation of horse-flies and deer-flies (Tabanidae: Diptera), II. The role of some visual factors in the attractiveness of decoy silhouettes.

Bracken, G. K., and Thorsteinson, A. J. (1965). *Entomogloia exp. appl.* **8**, 314–318. The orientation behaviour of horse-flies and deer-flies (Tabanidae: Diptera), IV. The influence of some physical modifications of visual decoys on orientation of horse-flies.

Breeland, S. G., and Eugene Pickard, (1965). *Mosquito News.* **25**, 19–21. The Malaise trap — an efficient and unbiased mosquito collecting device.

Broadbent, L. (1948). *Ann. app. Biol.* **35**, 379–394. Aphis migration and the efficiency of the trapping method.

Broadbent, L., Doncaster, J. P., Hull, R., and Watson, M. A. (1948). *Proc. R. ent. Soc. Lond.* **23**, 57–58. Equipment used for trapping and identifying alate aphides.

Broadbent, L., and Heathcote, G. D. (1961). *Entomologia exp. appl.* **4**, 226–237. Winged aphides trapped in potato fields 1942–1959.

Brook Worth, C., Paterson, H. R., and de Meillon, B. (1961). *Am. J. trop. Med. Hyg.* **10**, 583–592. The incidence of arthropod-borne viruses in a population of culicine mosquitoes in Tongaland, Union of South Africa (Jan., 1956 through April, 1960).

Bruce-Chwatt, L. J., Garret-Jones, C., and Weitz, B. (1966). *Bull. W.H.O.* **35**, 405–439. Ten years' study (1955–64) of host selection by anopheline mosquitoes.

Buescher, E. L., Scherer, W. F., Rosenberg, M. Z., Gresser, L., Hardy, J. L., and Bullock, H. R. (1959). *Am. J. trop. Med. Hyg.* **8**, 651–664. Ecological studies of Japanese encephalitis virus in Japan. II. Mosquito infection.

Burnett, G. F. (1960). *J. trop. Med. Hyg.* **63·** 153–162, 184–192, 208–215. Filariasis research in Fiji (1957–59).

Burnett, G. F. (1961a). *Nature, Lond.* **192**, 4798, 188. Effect of age and pregnancy on the tolerance of tse-tse flies to insecticides.

Burnett, G. F. (1961b). *Bull. ent. Res.* **52**, 531–539, 763–768. The susceptibility of tse-tse flies to topical applications of insecticides. Parts I and II.

Burnett, G. F. (1962). *Bull. ent. Res.* **53**, 337–354, 747–761. The susceptibility of tse-tse flies to topical applications of insecticides. Parts III–VI.

Burnett, G. F., Yeo, D., Miller, A. W. D., and White, P. J. (1961). *Bull. ent. Res.* **52**, 305–316. Aircraft application of insecticide in East Africa, XIII. An economical method for the control of *Glossina morsitans* Westw.

Burnett, G. F., Chadwick, P. R., Miller, A. W. D., and Beesley, J. S. S. (1965). *I.S.C.T.R. 10th Meeting, Kampala* (1964). 107–124. Aerial applications of insecticides in East Africa, XIV. Very low volume aerosol applications of Dieldrin and Telodrin for the control of *G. morsitans* Westw.

Bursell, E. (1961). *Proc. R. ent. Soc. Lond.* **36**, 9–20. The behaviour of tse-tse flies (*Glossina swynnertoni* Aust.) in relation to problems of sampling.

Bursell, E. (1966). *Bull. ent. Res.* **57**, 171–180. The nutritional state of tse-tse flies from different vegetation types in Rhodesia.

Buxton, P. A. (1955). *Mem. Lond. Sch. Hyg. trop. Med.* No. **10**, 816 pp. Lewis, London. The Natural History of Tse-tse Flies.

Carrion, A. L. (1930). *Publ. Hlth. Rep.* **45**, 1515–1520. Third report on a rat-flea survey of the city of San Juan, Porto Rico.

Chadwick, P. R. (1964). *Bull. ent. Res.* **55**, 23–28. A study of the resting sites of *Glossina swynnertoni* Aust. in northern Tanganyika.

Chadwick, P. R., Beesley, J. S. S., White, P. J., and Matechi, H. T. (1965). *I.S.C.T.R. 10th Meeting, Kampala* (1964), 97–106. An experiment on the eradication of *G. swynnertoni* Aust. by insecticidal treatment of the resting sites.

Challier, A. (1965). *Bull. Soc. Path. Exot.* **58**, 250–259. Amélioration de la méthode de détermination de l'age physiologique des Glossines. Etudes faites sur *Glossina palpalis gambiensis* Vanderplank., 1949.

Chow, C. Y., and Thevasagayan, E. S. (1957). *Bull. W.H.O.* **16**, 609–632. Bionomics and control of *Culex pipiens fatigans* Wied. in Ceylon.

Clark, L. R., Geier, P. W., Hughes, R. D., and Morris, R. F. (1967). *The Ecology of Insect Populations in Theory and Practice*, 232 pp. Methuen and Co. Ltd., London.

Cole, L. C. (1945). *Publ. Health. Rep. Wash.* **60**, 45, 1337–1342. The effect of temperature on the sex ratio of *Xenopsylla cheopis* recovered from live rats.

Colless, D. H. (1959). *Ann. trop. Med. Parasit.* **53**, 251–258. Notes on the culicine fauna of Singapore, VI. Observations on catches made with baited and unbaited trap nets.

Colless, D. H. (1959b). *Ann. trop. Med. Parasit.* **53**, 259–267. Notes on the culicine fauna of Singapore, VII. Host preference in relation to transmission of the disease.

Corbet, P. S. (1961). *Trans. R. Soc. ent. Lond.* **113**, 301–314. Entomological studies from a high tower in Mpanga forest, Uganda, VI. Nocturnal flight activity of Culicidae and Tabanidae as indicated by light traps.

Corbet, P. S. (1964). *Proc. R. ent. Soc. Lond.* **39**, 53–67. Nocturnal flight activity of sylvan Culicidae and Tabanidae (Diptera) as indicated by light traps: a further study.

Coz, J., Eyraud, M., Venard, P., Attiou, B., Somda, D., and Ouedraogo, V. (1965). *Bull. W.H.O.* **33**, 435–452. Expériences en Haute-Volta sur l'utilisation de cases pièges pour la mesure de l'activité du D.D.T. contre les moustiques.

Crisp, G. (1956). *Simulium and onchocerciasis in the Northern Territories of the Gold Coast*, 171 pp. H. K. Lewis, London.

Crosskey, R. W. (1958). *Bull. ent. Res.* **49**, 715–735. First results in the control of *Simulium damnosum* Theobald (*Diptera, Simuliidae*) in northern Nigeria.

Dalmat, H. T. (1955). *Smithson. misc. Collns.* **125**, No. 1. 425 pp. The black-flies (*Diptera, Simuliidae*) of Guatemala and their role as vectors of onchocerciasis.

Dame, D. A., and Fye, R. L. (1964). *J. econ. Ent.* **57**, 776–777. Studies on feeding behaviour of house-flies.

Dame, D. A., Deane, G. J. W., and Ford, J. (1965). *I.S.C.T.R. 10th Meeting, Kampala, 1964. Publ. Comm. tech. Co-op. Afr.* No. **97**, 93–96. Investigations of the sterile male technique with *Glossina morsitans* Westw.

Davidson, A. (1962). *Riv. Parassit.* **23**, 61–70. Trapping house-flies in the rural areas of Israel.

Davidson, G. (1953). *Bull. ent. Res.* **44**, 231–254. Experiments on the effect of residual insecticides in houses against *Anopheles gambiae* and *A. funestus*.

Davies, J. B., Crosskey, R. W., Johnston, M. R. L., and Crosskey, M. E. (1962). *Bull. W.H.O.* **27**, 491–510. The control of *Simulium damnosum* at Abuja, Northern Nigeria, 1955–60.

Davies, L. (1957a). *Bull. ent. Res.* **48**, 407–424. A study of the black-fly, *Simulium ornatum* Mg. (Diptera) with particular reference to its activity on grazing cattle.

Davies, L. (1957b). *Bull. ent. Res.* **48**, 535–552. A study of the age of females of *Simulium ornatum* Mg. (Diptera) attracted to cattle.

Davies, L. (1965). *W.H.O. Oncho/Inf/3.65.* A case for conducting light-trap studies on African vector *Simulium.*

Davies, L., and Williams, C. B. (1962). *Trans. R. ent. Soc. Lond.* **116·** 1–20. Studies on black-flies (*Diptera: Simuliidae*) taken in a light trap in Scotland. Part I. Seasonal distribution, sex ratio, and internal condition of catches.

Davies, L., Downe, A. E. R., Weitz, B., and Williams, C. B. (1962). *Trans. R. ent. Soc. Lond.* **114·** 21–27. Studies on black-flies (*Diptera: Simuliidae*) taken in a light trap in Scotland, II. Blood-meal identification by precipitin tests.

Davis, D. H. S. (1939). *S. Afr. J. Sci.* **36**, 438. Some ecological methods in research on bubonic plague.

Davis, D. H. S. (1953). *Bull. W.H.O.* **9**, 665–700. Plague in Africa from 1935 to 1949. A survey of wild rodents in African territories.

Davis, D. H. S. (1964). *Ecological studies in Southern Africa*, 415 pp. "Ecology of Wild Rodent Plague", 301–314.

Deane, L. M., Damasceno, R. G., and Arouck, R. (1953). *Folia. clin. biol.* **20**, 2, 101–110. Vertical distribution of mosquitoes in a forest near Belem, Para.

Deguisti, D. L., and Hartley, C. F. (1965). *Am. J. trop. Med. Hyg.* **14**, 309–313. Ectoparasites of rats from Detroit, Michigan.

Detinova, T. S. (1962). Age-grouping methods in Diptera of Medical Importance. W.H.O. Geneva. Monograph Series. No. **47**, 216 pp.

Detinova, T. S., and Gillies, M. T. (1964). *Bull. W.H.O.* **30**, 23–28. Observations on the determination of the age composition and epidemiological importance of populations of *Anopheles gambiae* Giles and *Anopheles funestus* Giles, in Tanganyika.

de Foliart, G. R., and Rao, M. Ramachandra (1965). *J. med. Ent.* **2**, 84–85. The ornithophilic black-fly *Simulium meridionale* Riley (*Diptera: Simuliidae*) feeding on man during autumn.

de Meillon, B., Anthony Sebastian and Z. H. Khan (1967). *Bull. W.H.O.* **36**, 53–65. Cane sugar feeding in *Culex pipiens fatigans.*

de Meillon, B., Anthony Sebastian and Z. H. Khan (1967). *Bull. W.H.O.* **36**, 39–46. Time of arrival of gravid *Culex pipiens fatigans* at an ovipositing site, the oviposition cycle and the relationship between time of feeding and time of oviposition.

de Meillon, B., Anthony Sebastian and Z. H. Khan (1966). *Bull. W.H.O.* **35**, 808–809. Positive geotaxis in gravid *Culex pipiens fatigans.*

de Meillon, B., and Anthony Sebastian. (1967). *Bull. W.H.O.* **36·** 75–80. Qualitative and quantitative characteristics of adult *Culex pipiens fatigans* populations according to time, site and place of capture.

de Meillon, B., and Z. H. Khan (1967). *Bull. W.H.O.* **36**, 169–174. Examples of the use of simple age-grading in the assessment of *Culex pipiens fatigans* populations.

de Meillon, B., Myo Paing., Anthony Sebastian, and Z. H. Khan (1967). *Bull. W.H.O.* **36**, 67–73. Outdoor resting of *Culex pipiens fatigans* in Rangoon, Burma.

de Meillon, B., Paterson, H. E., and Muspratt, J. (1957). *S. Afr. J. med. Res.* **22**, 47–53. Studies on arthropod-borne viruses of Tongaland, II. Notes on the more common mosquitoes.

de Zulueta, J. (1950). *Nature, Lond.* **166**, 180. Biology of adult mosquitoes in Eastern Colombia.

de Zulueta, J. (1952). *Am. J. trop. Med. Hyg.* **1**, 314–329. Observations on mosquito density in an endemic malarious area in Eastern Colombia.

de Zulueta, J. (1964). *Riv. Malar.* **43**, 29–36. Ethological changes in malaria vectors. A review of the situation in the light of recent findings.

Disney, R. H. L. (1966). *Bull. ent. Res.* **56**, 445–451. A trap for phlebotomine sand-flies attracted to rats.

Dow, R. P., Reeves, W. P., and Bellamy, R. E. (1957). *Am. J. trop. Med. Hyg.* **6**, 294–303. Field tests of avian host preference of *Culex tarsalis* Coq.

Downe, A. E. R. (1962). *Can. J. Zool.* **40**, 725–732. Some aspects of host selectivity by *Mansonia perturbans* Walk *(Diptera: Culicidae)*.

Duke, B. O. L. (1959a). *Ann. trop. Med. Parasit.* **52**, 24. Studies on the biting habits of *Chrysops*, Part V.

Duke, B. O. L. (1959b). *Ann. trop. Med. Parasit.* **53**, 203–214. Studies on the biting habits of *Chrysops*, Part VI.

Duke, B. O. L., Lewis, D. J., and Moore, P. J. (1966). *Ann. trop. Med. Parasit.* **60**, 3, 318–336. Onchocerca-*Simulium* complexes. I. Transmission of forest and Sudan-Savannah strains of *Onchocerca volvulus*, from Cameroon, by *Simulium damnosum* from various West African bioclimatic zones.

Dunbar, R. W. (1966). *Nature,* **5023**, 597–599. Four sibling species included in *Simulium damnosum* Theobald *(Diptera: Simuliidae)* from Uganda.

Dyce, A. L., and Lee, D. J. (1962). *Australian. J. Zool.* **10**, 84–94. Blood-sucking flies (Diptera) and myxomatosis transmission in a mountain environment in South Wales.

Eskey, C. R. (1930). *Publ. Hlth. Reps. Wash.* **45**, No. 36, 2077–2115, No. 37, 2162–2187. Chief etiological factors of plague in Ecuador and the anti-plague campaign.

Eskey, C. R. (1934). *Publ. Hlth. Bull. Wash.* No. **213**. 70 pp. U.S. Public Health Service. Epidemiological study of plague in the Hawaiian Islands.

Eskey, C. R., and Haas, V. H. (1940). *Publ. Hlth. Bull. Wash.* No. **254**. 83 pp. Plague in the western part of the United States.

Estrade, F. (1934). *Bull. Soc. Path. Exot.* **27**, 458–461. Technique et appareil pour la capture des puces dans les poussières et débris de céréales.

Fallis, A. M. (1964). *Exp. Parasit.* **15**, 439–447. Feeding and related behaviour of female *Simuliidae* (Diptera).

Fallis, A. M., and Smith, S. M. (1964). *Can. J. Zool.* **42**, 723–730. Ether extracts from birds and CO_2 as attractants for some ornithophilic simuliids.

Ford, J. (1962). Microclimates of tse-tse fly resting sites in the Zambesi valley, Southern Rhodesia, I.S.C.T.R. 9th meeting. Publ. No. 88. Conakry, 165–170.

Ford, J., Glasgow, J. P., Johns, D. L., and Welch, J. R. (1959). *Bull. ent. Res.* **50**, 275–285. Transect fly rounds in field studies of *Glossina*.

Foster, R. (1964). *Bull. ent. Res.* **54**, 727–744. Contribution to the epidemiology of human sleeping sickness in Liberia: Bionomics of the vector *Glossina palpalis* (R-D) in a forest habitat.

Foster, R., White, P. J., and Yeo, D. (1961). *Bull. ent. Res.* **52**, 293–303. Aircraft application of insecticides in East Africa, XII.

Fox, I. (1953). *J. econ. Ent.* **45**, 888–889. Light trap studies on *Culicoides* in Puerto Rico.

Fox, I. (1956). *Am. J. trop. Med. Hyg.* **5**, 893–900. Murine typhus fever and rat ectoparasites in Puerto Rico.

Fox, I. (1958). *Mosquito News.* **18**, 117. The mosquitoes of the international airport Isla Verde, Puerto Rico, as shown by light traps.

Fox, I., and Kohler, C. E. (1950). *Puerto Rico. J. publ. Hlth. trop. Med.* **25**, 342–349. Distribution and relative abundance of the species of biting midges or *Culicoides* in eastern Puerto Rico as shown by light traps.

Fox, I., and Ileana Garcia-Moll (1961). *Am. J. trop. Med. Hyg.* **10**, 566–573. Rat ectoparasite surveys in relation to murine typhus fever in Puerto Rico.

Fox, R. M. (1957). *Am. J. trop. Med. Hyg.* **6**, 598–620. Anopheles gambiae in relation to malaria and filariasis in coastal Liberia.

Fredeen, F. J. H. (1961). *Can. Ent.* **93**, 73–78. A trap for studying the attacking behaviour of black-flies, *Simulium arcticum* Ma.

Fredeen, F. J. H., Spinks, J. W. T., Anderson, J. R., Arnason, A. P., and Rempel, J. G. (1953). *Can. J. Zool.* **31**, 1–15. Mass tagging of black-flies (*Diptera : Simuliidae*) with radiophosphorus.

Galindo, P., Trapido, H., and Carpenter, S. J. (1950). *Am. J. trop. Med. Hyg.* **30**, 533–574. Observations on diurnal forest mosquitoes in relation to sylvan yellow fever in Panama.

Gelfand, H. M. (1955). *Trans. R. Soc. Trop. Med. Hyg.* **49**, 508–527. Anopheles gambiae Giles and *A. melas* Theobald in a coastal area of Liberia, West Africa.

George, P. V., and Webster, W. J. (1934). *Ind. J. med. Res.* **22**, 77–104. Plague enquiry in the Cumbum valley, South India.

Giglioli, G. (1948). *Malaria, Filariasis and Yellow Fever in British Guiana.* 226 pp. Mosquito Control Service, Medical Dept. British Guiana.

Gillies, M. T. (1954). *Bull. ent. Res.* **45**, 361–387. Studies on house leaving and outside resting of *Anopheles gambiae*, Giles and *A. funestus* Giles in East Africa: I, The outside-resting population; and II, The exodus from houses and the house-resting population.

Gillies, M. T. (1955). *Am. J. trop. Med. Hyg.* **4**, 1103–1113. The density of adult *Anopheles* in the neighbourhood of an East African village.

Gillies, M. T. (1961). *Bull. ent. Res.* **52**, 99–127. Studies on the dispersion and survival of *Anopheles gambiae* Giles in East Africa, by means of marking and release experiments.

Gillies, M. T., and Wilkes, T. J. (1963). *Ann. trop. Med. Parasit.* **57**, 204–213. Observations on nulliparous and parous rates in a population of *Anopheles funestus* in East Africa.

Gillies, M. T., and Wilkes, T. J. (1965). *Bull. ent. Res.* **56**, 237–262. A study of the age composition of *Anopheles gambiae* Giles and *A. funestus* Giles in north-eastern Tanzania.

Girard, G. (1943). *Bull. Soc. Path. Exot.* **36**, 4–41. Les ectoparasites de l'homme dans l'épidemologie de la pest.

Glasgow, J. P. (1961). *Bull. ent. Res.* **51**, 781–785. The variability of fly-round catches in field studies of *Glossina*.

Glasgow, J. P. (1963). *The distribution and abundance of tse-tse.* 241 pp. Pergamon Press, Oxford.

Glasgow, J. P. (1967). *Ann. Rev. Ent.* **12**, 421–438. Recent fundamental work on tse-tse flies.

Glasgow, J. P., Isherwood, F., Lee-Jones, F., and Weitz, B. (1958). *J. anim. Ecol.* **27**, 59–69. Factors influencing the staple food of tse-tse flies.

Glover, P. E., Jackson, C. H. N., Robertson, A. G., and Thompson, W. E. F. (1955). *Bull. ent. Res.* **46**, 57–67. The extermination of the tse-tse fly *Glossina morsitans* Westw. at Abercorn, Northern Rhodesia.

Goma, L. K. H. (1965). *Bull. ent. Res.* **56**· 17–35. The flight activity of some East African mosquitoes (*Diptera : Culicidae*), I. Studies on a high tower in Zika forest, Uganda.

Greenberg, B. (1964). *Am. J. Hyg.* **80**, 149–156. Experimental transmission of *Salmonella typhimurium* by house-flies to man.

Haas, G. E. (1966). *J. med. Ent.* **24**, 392–395. A technique for estimating the total number of rodent fleas in cane fields in Hawaii.

Haddow, A. J. (1942). *Bull. ent. Res.* **33**, 91–142. The mosquito fauna and climate of native huts at Kisumu, Kenya.

Haddow, A. J. (1954). *Bull. ent. Res.* **45**, 199–242. Studies on the biting habits of African mosquitoes; and appraisal of methods employed with special reference to the 24-hour catch.

Haddow, A. J. (1961). *Bull. ent. Res.* **52**, 317–351. Studies on the biting habits and medical importance of East African mosquitoes of the genus Aedes, II subgenera Mucida, Diceromyia, Finlaya and Stegomyia.

Haddow, A. J., Corbet, P. S., and Gillett, J. D. (1961). *Trans. R. ent. Soc. Lond.* **113**, 249–368. Entomological studies from a high tower in Mpanga Forest, Uganda.

Hadjinicolau, J. (1963), see *W.H.O.* (1963).

Hafez, M., and Attia, M. A. (1958). *Bull. Soc. ent. Egypt.* **42·** 83–121. Studies on the ecology of *Musca sorbens* Wied. in Egypt.

Hanec, W., and Bracken, G. K. (1964). *Can. Ent.* **96**, 136, 201, 369. Seasonal and geographic distribution of Tabanidae (Diptera) in Manitoba, based on females captured in traps.

Harley, J. M. B. (1965). *Bull. ent. Res.* **56**, 141–160. Activity cycles of *Glossina pallidipes* Aust. *G. palpalis fuscipes* Newst., and *G. brevipalpis* Newst.

Harley, J. M. B. (1966). *Bull. ent. Res.* **57**, 23–37. Studies on age and trypanosome infection rate in females of *Glossina pallidipes* Aust., *G. palpalis fuscipes* Newst., and *G. brevipalpis* Newst.

Harley, J. M. B. (1967a). *Bull. ent. Res.* **57**, 459–477. Further studies on age and trypanosome infection rates in *Glossina pallidipes* Aust., *G. palpalis fuscipes* Newst., and *G. brevipalpis* Newst. in Uganda.

Harley, J. M. B. (1967b). *Entomologia expt. appl.* **10**, 240–252. The influence of sampling method on the trypanosome infection rates of catches of *Glossina pallidipes* and *G. fuscipes*.

Hayes, R. O., Bellamy, R. E., Reeves, W. E., and Willis, M. J. (1958). *Mosquito News.* **18**, 218–227. Comparison of 4 sampling methods for measurement of *Culex tarsalis* adult populations.

Heisch, R. B., Grainger, W. E., and d'Souza, J. S. A. M. (1953). *Trans. R. Soc. trop. Med. Hyg.* **47**, 503–521. Results of a plague investigation in Kenya.

Hirst, L. F. (1933). *Ceylon, J. Sci.* **3**, 49–113. A rat-flea survey of Ceylon with a brief discussion of recent work on rat-flea species distribution in relation to the spread of bubonic plague in the East Indies.

Hirst, L. F. (1953). *The conquest of plague.* 478 pp. Clarendon Press, Oxford.

Holway, R. T., Mitchell, W. A., and Abdel Aziz Salah (1951). *An. ent. Soc. Am.* **44**, 381–398. Studies on the seasonal prevalence and dispersal of the Egyptian housefly: I. The adult fly.

Hoogstraal, H., Dietlein, D. R., and Heyneman, D. (1962). *Trans. R. Soc. trop. Med. Hyg.* **56**, 411–422. Leishmaniasis in the Sudan Republic: 4. Preliminary observations on man-biting sand-flies (*Psychodidae: Phlebotomus*) in certain upper Nile endemic areas.

Hu, S. M. K., and Grayson, J. T. (1962). *Am. J. trop. Med. Hyg.* **11**, 131–140. Encephalitis in Taiwan: II. Mosquito collection and bionomic studies.

Hurlbut, H. S., and Weitz, B. (1956). *Am. J. trop. Med. Hyg.* **5**, 901–908. Some observations on the bionomics of the common mosquitoes of the Nile delta.

Isherwood, F. (1957). *Bull. ent. Res.* **48**, 601–606. The resting sites of *Glossina swynnertoni* Aust. in the wet season.

152   ECOLOGY OF INSECT VECTOR POPULATIONS

Jackson, C. H. N. (1944). *Ann. Eugen.* **12**, 176–205. The analysis of a tse-tse fly population, II.

Jackson, C. H. N. (1947). *Ann. Eugen.* **14**, 91–108. The analysis of a tse-tse fly population, III.

Jackson, C. H. N. (1949). *Biol. Rev.* **24**, 174–199. The biology of tse-tse flies.

Jackson, C. H. N. (1953). *J. Anim. Ecol.* **22**, 78–86. A mixed population of *Glossina morsitans* and *G. swynnertoni*.

Jackson, C. H. N. (1954). *J. Anim. Ecol.* **23**, 368–372. The hunger cycle of *Glossina morsitans* Westwood and *G. swynnertoni* Aust.

Jamnback, H., and Watthews, T. (1963). *Ann. ent. Soc. Am.* **56**, 728–732. Studies of populations of adult and immature *Culicoides sanguisuga* (*Diptera: Ceratopogonidae*)

Jewell, G. R. (1956). *Nature,* **178**, No. 4536, 750. Marking of tse-tse flies for their detection at night.

Jewell, G. R. (1958). *Nature,* **181**. No. 4619, 1354. Detection of tse-tse fly at night.

Johnston, C. G., and Taylor, L. R. (1955). *Ann. appl. Biol.* **43**, 51–62. The development of large suction traps for airborne insects.

Joseph, C., Menon, M. A. U., and Nair, G. K. (1962). *Ind. J. Mal.* **14**, 663–686. Pilot studies on the control of filariasis due to *Brugia malayi* (Brug. 1927), in Kerala.

Jordan, A. M. (1965). *Bull. ent. Res.* **56**, 1–16. Observations on the ecology of *Glossina morsitans submorsitans* Newst. in the northern Guinea savannah of northern Nigeria.

Jordan, A. M., Lee-Jones, F., and Weitz, B. (1961). *Ann. trop. Med. Parasit.* **55**, 167–179. The natural hosts of tse-tse flies in the forest belt of Nigeria and the southern Cameroons.

Jordan, A. M., Page, W. A., and McDonald, W. A. (1960). *Int. Sci. Comm. for Tryps. Res.* (*I.S.C.T.R.*) *Publ. Comm. tech. Co-op. Afr. S.* No. **41**. 315–317. Progress made in ascertaining the natural hosts favoured by different species of tse-tse.

Kartman, L. (1946). *J. Parasit.* **23**, 30–35. A note on the problem of plague in Dakar, Senegal, French West Africa.

Kernaghan, R. J. (1961). *J. trop. Med. Hyg.* **64**, 303–309. Insecticidal control of the vectors of human trypanosomiasis in Northern Nigeria.

Kettle, D. S. (1960). *Bull. ent. Res.* **51**, 461–489. The flight of *Culicoides impunctatus* Goetghbuer (*Diptera: Ceratopogonidae*) over moorland and its bearing on midge control.

Kettle, D. S. (1962). *Ann. Rev. Ent.* **7**, 401–408. The bionomics and control of *Culicoides* and *Leptoconops* (*Diptera: Ceratopogonidae Heleidae*).

King, H. H., Iyer, P. V. S., Natarajan, N., and George, P. V. (1929). *Ind. J. med. Res.* **17**, 297–334. A rat flea survey of the Madras Presidency.

Kirk, R., and Lewis, D. J. (1947). *Trans. R. Soc. trop. Med. Hyg.* **40**, 869–888. Studies in leishmaniasis in the Anglo-Egyptian Sudan, IX. Further observations on the sand-flies (*Phlebotomus*) of the Sudan.

Kirk, R., and Lewis, D. J. (1951). *Trans. R. Soc. trop. Med. Hyg.* **102**, 383–510. The Phlebotominae of the Ethiopian Region.

Klein, J. M. (1963). *Bull. Soc. Path. Exot.* **56**, 1202–1230. Ecological observations on the fleas of Meriones spp. in a natural focus of plague in Kurdistan.

Knight, K. L. (1964). *J. med. Ent.* **1**, 109–115. Quantitative methods for mosquito larval surveys.

La Croix, E. A. S. (1960). *C.C.T.A. 8th Meeting, Jos, Nigeria.* 65–76. Some entomological aspects of a series of outbreaks of human trypanosomiasis in Northern Ghana in 1957–59.

Lambrecht, F. L. (1958). *Folia scient. Afr. cent.* IV. No. **1**. Une nouvelle technique pour l'etude ecologique des Glossines de savane.

Lamontellerie, M. (1963). *Bull. de l'I.F.A.N.* **25**, 467–484. Observations sur *Simulium adersi* Pomeroy en zone de savane seche (Region de Garango, Haute-Volta).

Lamontellerie, M. (1967). *Bull. I.F.A.N.* *29*, No. **4**. 1812–1832. Captures de Dipteres Simuliidae de nuit en zone de savane seche.

Langridge, W. P., Kernaghan, R. J., and Glover, P. E. (1963). *Bull. W.H.O.* **28**, 671–701. A review of recent knowledge of the ecology of the main vectors of trypanosomiasis.

Laurence, B. R. (1963). *Bull. W.H.O.* **28**, 229–234. Natural mortality in two filarial vectors.

Laurence, B. R. (1966). *Bull. W.H.O.* **34**, 475–477. Predation in a mosquito community.

Le Berre, R. (1966). *Memoires O.R.S.T.O.M.* No. **17**, 204 pp. Contribution a l'etude biologique et ecologique de *Simulium damnosum* Theobald, 1903 (*Diptera, Simuliidae*) 8 vo. Paris. Office de la Recherche Scientifique et Technique Outre-Mer.

Le Berre, R., Balay, G., Bringues, J., and Coz, J. (1964). *Bull. W.H.O.* **31**, 843–855. Biologie et ecologie de la femelle de *Simulium damnosum* Theobald, 1903, en fonction des zones bioclimatiques d'Afrique occidentale.

Leggate, B. M., and Pilson, R. D. (1961). *Bull. ent. Res.* **51**, 697–704. The diurnal feeding activity of *Glossina pallidipes* Aust. in relation to trypanosome challenge.

Lewis, D. J. (1956). *Proc. 10th Int. Congress. Ent. Montreal*, **3**, 541–550. *Simulium damnosum* in the Tonkolili Valley, Sierra Leone.

Lewis, D. J. (1960). *Ann. trop. Med. Parasit.* **54**, 208–223. Observations on *Simulium damnosum* in the Southern Cameroons and Liberia.

Lewis, D. J., and Duke, B. O. L. (1966). *Ann. trop. Med. Parasit.* **60**, 337–346. Onchocerca-Simulium complexes, II. Variation in West African female *Simulium damnosum*.

Lewis, D. J., and Cheryl R. Domoney (1966). *Proc. R. ent. Soc. Lond.* A**41**, 175–179. Sugar meals in *Phlebotominae* and *Simuliidae* (Diptera).

Lewis, D. J., Lyons, G. R. L., and Marr, J. D. M. (1961). *Ann. trop. Med. Parasit.* **55**, 202–210. Observations on *Simulium damnosum* from the Red Volta in Ghana.

Lewis, D. J., and Murphy, D. H. (1965). *J. med. Ent.* **1**, 371–376. The sand-flies of the Gambia (*Diptera: Phlebotomiinae*).

Lindquist, A. W., Ikeshoji, B. Grab., de Meillon, B., and Z. H. Khan (1967). *Bull. W.H.O.* **36**, 21–37. Dispersion studies of *Culex pipiens fatigans* tagged with ^{32}P in the Kemmendine area of Rangoon, Burma.

Lindsay, D. R., and Scudder, H. I. (1956). *Ann. Rev. Ent.* **1**, 323–346. Non-biting flies and disease.

Lockley, R. M. (1954). *Vet. Rec.* **66**, 434. The European Rabbit flea, *Spilopsyllus cuniculi*, as a vector of myxomatosis in Britain.

Loomis, E. C., and Myers, E. G. (1960). *Am. J. Hyg.* **71**, 378–388. California encephalitis surveillance program. Mosquito population measurement and ecologic consideration.

Loomis, E. C., and Sherman, E. J. (1959). *Mosquito News.* **19**, 232–237. Comparison of artificial shelter and light traps for measurement of *Culex tarsalis* and *Anopheles freeborni* populations.

Love, G. J., and Smith, W. W. (1957). *Mosquito News.* **17**, 9–14. Preliminary observations on the relation of light-trap collections to mechanical sweep-net collections in sampling mosquito populations.

Love, G. J., and Smith, W. W. (1958). *Mosquito News.* **18**, 279–283. The stratification of mosquitoes.

Lumsden, W. H. R. (1951). *Bull. ent. Res.* **42**, 317–330. Probable insect vectors of yellow fever virus from monkey to man in Bwamba county, Uganda.

Lumsden, W. H. R. (1957). *Bull. ent. Res.* **48**, 769–782. The activity cycle of domestic *Aedes aegypti* (L) (*Diptera: Culicidae*) in Southern Province, Tanganyika.

Lumsden, W. H. R. (1958). *Nature*, **181·** 819–820. A trap for insects biting small vertebrates.

Macchiavello, A. (1955). Practical absolute rat-flea index used in plague control, pp. 24–27. In *Expert Committee on Plague*. World Health Organization, Technical Report Series, No. 11.

Madwar, S., and Zahar, A. R. (1951). *Bull. W.H.O.* **3**, 621–636. Preliminary studies on house-flies in Egypt.

Mahood, A. R. (1962). *I.S.C.T.R., 9th meeting, Conakry*, 181–185. A note on the ecology of *Glossina morsitans submorsitans* Newst. in the Guinea Savannah zone of Northern Nigeria.

Marr, J. D. M. (1962). *Bull. W.H.O.* **27**, 622–847. The use of an artificial breeding site and cage in the study of *Simulium damnosum* Theobald.

Marr, J. D. M., and Lewis, D. J. (1964). *Bull. ent. Res.* **55**, 547–564. Observations on the dry season survival of *Simulium damnosum* Theo. in Ghana.

McDonald, W. A. (1960a). *Nature*. **185**, 867–868. Nocturnal detection of tse-tse flies in Nigeria with ultra-violet light.

McDonald, W. A. (1960b). *I.S.C.T.R. 8th Meeting, Jos, Nigeria*. 243–245. Insecticidal spraying against *Glossina palpalis* in Nigeria based on a study of the nocturnal resting sites with ultra-violet light.

MacLeod, J. (1958). *Trans. R. ent. Soc. Lond.* **110**, 363–392. The estimation of numbers of mobile insects from low incidence recapture data.

MacLeod, J., and Donnelly, J. (1957a). *J. Anim. Ecol.* **26**, 135–170. Some ecological relationships of natural populations of calliphorine blow-flies.

Macleod, J., and Donnelly, J. (1957b). *Bull. ent. Res.* **48**, 585–592. Individual and group marking methods for fly-population studies.

Macleod, J., and Donnelly, J. (1962). *J. Anim. Ecol.* **31**, 525–543. Microgeographic aggregation in blow-fly populations.

McMahon, J. P., Highton, R. B., and Goiny, H. (1958). *Bull. W.H.O.*, **19**, 71–81. The eradication of *Simulium neavei* from Kenya.

McRae, A. W. R. (1967). The *Simulium damnosum* species complex. E. Afr. Virus Research Inst. Report 1966, 38–39. E. Afr. Common. Services.

Mead-Briggs, A. R. (1964a). *J. expl. Biol.* **41**, 371–402. The reproductive biology of the rabbit flea *Spilopsyllus cuniculi* (Dale) and the dependence of this species upon the breeding of its host.

Mead-Briggs, A. R. (1964b). *Ent. Month. Mag.* **100**, 8–17. Records of rabbit fleas *Spilopsyllus cuniculi* from every county in Great Britain with notes on infestation rates.

Mead-Briggs, A. R. (1964c). *J. Anim. Ecol.* **33**, 13–26. Some experiences concerning the interchange of rabbit fleas, *Spilopsyllus cuniculi* (Dale) between living rabbit hosts.

Mead-Briggs, A. R and Rudge, A. J. B. (1960). Breeding of the Rabbit flea, *Spilopsyllus cuniculi* (Dale): Requirement of a 'factor' from a pregnant rabbit for ovarian maturation. *Nature*. **187**, 1136–1137.

Miles, V. I., Kenney, A. R., and Stark, H. E. (1957). *Am. J. Trop. Med. Hyg.* **6**, 752. Flea-host relationship of associated *Rattus* and native wild rodents in the San Francisco Bay area of California with special reference to plague.

Minter, D. M. (1961). *Bull. ent. Res.* **52**, 233–238. A modified Lumsden suction trap for biting insects.

Minter, D. M. (1964). *Bull. ent. Res.* **55**, 421–435. Seasonal changes in population of phlebotomine sand-flies (*Diptera: Psychodidae*) in Kenya.

Mironov, N. P. *et al.* [In Russian] (1963). *Zool. Zh.* **3**, 384–394. The spatial distribution of fleas in burrows of *Citellus pygmaeus* and the rationalization of methods for assessing their numbers.

Mitchell, J. A., Pirie, J. H. H., and Ingram, A. (1927). *Publ. S. Afr. Inst. Med. Res.* No. **20**, vol. iii, 85–256. The plague problem in South Africa: Historical, bacteriological and entomological studies.

Mohr, C. O. (1951). *Am. J. trop. Med.* **31**, 355–372. Entomological background and the distribution of murine typhus and murine plague in the United States.

Mohr, C. O., and Lord, R. D. (1960). *J. Wildlife Management.* **24**, 290–297. Relation of ectoparasite populations to rabbit populations in northern Illinois.

Moorhouse, D. E. and Wharton, R. H. (1965). *J. med. Ent.* **1**, 359–370. Studies on Malayan vectors of malaria; methods of trapping and observations on biting cycles.

Morris, K. R. S. (1960). *Bull. ent. Res.* **51**, 533–557. Trapping as a means of studying the game tse-tse *Glossina pallidipes* Aust.

Morris, K. R. S. (1961a). *Bull. ent. Res.* **52**, 239–256. Problems in the assessment of tse-tse populations.

Morris, K. R. S. (1961b). *Am. J. trop. Med. Hyg.* **10**, 905–913. Effectiveness of traps in tse-tse surveys in the Liberian rain forest.

Morris, K. R. S. (1962). *J. trop. Med. Hyg.* **65**, 12–23. The food of *Glossina palpalis* (R–D) and its bearing on the control of sleeping sickness in forest country.

Morris, K. R. S., and Morris, M. G. (1949). *Bull. ent. Res.* **39**, 491–528. The use of traps against tse-tse in West Africa.

Morris, R. F. (1960). *Ann. Rev. Ent.* **5**, 243–264. Sampling insect populations.

Mourier, H. (1965). *Vidensk Meda, fra Dansk. naturh. Foren., bd.* **128**, 221–231. The behaviour of house-flies (*Musca domestica* L) towards new objects.

Muirhead-Thomson, R. C. (1951). *Mosquito Behaviour in relation to malaria transmission and control in the tropics.* pp. 219. London. Edward Arnold and Co.

Muirhead-Thomson, R. C. (1956). *J. Hyg.* **54**, 461–471. The role of woodland Aedes mosquitoes in the transmission of myxomatosis in England.

Muirhead-Thomson, R. C. (1956b). Observations on the European rabbit flea (*Spilopsyllus cuniculi*) in relation to myxomatosis in England. Unpublished report to the Infestation Control Division, Ministry of Agriculture, Fisheries and Food.

Muirhead-Thomson, R. C. (1956). *Nature.* **178**, 1297–1299. Communal oviposition in *Simulium damnosum* Theobald (*Diptera: Simuliidae*)

Muirhead-Thomson, R. C. (1958). *Bull. W.H.O.* **19**, 1116–1118. A pit shelter for sampling outdoor mosquito populations.

Muirhead-Thomson, R. C. (1960a). *W.H.O./Mal/276.* World Health Organization, Geneva. Further studies on the use of the artificial pit shelter for sampling outdoor resting populations of African anophelines.

Muirhead-Thomson, R. C. (1960b). *Bull. W.H.O.* **22**, 721–734. The significance of irritability, behaviouristic avoidance and allied phenomena in malaria eradication.

Muirhead-Thomson, R. C. (1963) see *W.H.O.*, **1963**.

Muirhead-Thomson, R. C., and Mercier, E. C. (1952). *Ann. trop. Med. Parasit.* **46**, 103–116; **48**, 201–213. Factors in malaria transmission in Jamaica, I and II.

Murrosh, C. M., and Thaggard, C. W. (1966). *Ann. ent. Soc. Am.* **59**, 533–547. Ecological studies of the house-fly.

156   ECOLOGY OF INSECT VECTOR POPULATIONS

Myers, K. (1956). *C.S.I.R.O. Wild Res.* **1**, 45–58. Methods of sampling winged insects feeding on the rabbit, *Oryctolagus cuniculus* (L).

Nash, T. A. M. (1952). *Bull. ent. Res.* **43**, 33–42. Some observations on resting tse-tse fly populations, and evidence that *Glossina medicorum* is a carrier of trypanosomes.

Nash, T. A. M. (1960). *Trop. Dis. Bull.* **57**, 973–1003. A review of the African trypanosomiasis problem.

Nash, T. A. M., and Davey, J. T. (1950). *Bull. ent. Res.* **41**, 153–157. The resting habits of *G. medicorum, G. fusca* and *G. longipalpis*.

Nash, T. A. M., and Page, W. A. (1953). *Trans. R. ent. Soc. Lond.* **104**, 71–169. The ecology of *Glossina palpalis* in Northern Nigeria.

Nash, T. A. M., and Steiner, J. O. (1957). *Bull. ent. Res.* **48**, 323–339. The effect of obstructive clearing on *Glossina palpalis* (R–D).

Nelson, R. L. (1966). *Proc. 34th Ann. Conference of the California Mosquito Control Assn.* California. 65–66. New collecting methods for vectors of arbiviruses.

Nelson, D. B., and Chamberlain, R. W. (1955). *Mosquito News.* **15**, 28–32. A light trap and mechanical aspirator operated on dry cell batteries.

Nicholas, W. L., Kershaw, W. E., Keay, R. W. J., and Zahra, A. (1953). *Ann. trop. Med. Parasit.* **47**, 95–111. Studies on the epidemiology of filariasis in West Africa with special reference to the British Cameroons and the Niger Delta: III. The distribution of *Culicoides* spp. biting man in the rain forest, the forest fringe and the mountain grasslands of the British Cameroons.

Nielsen, E. T., and Nielsen, A. T. (1953). *Ecology*, **34**, 141–156. Field observations on the habits of *Aedes taeniorhynchus*.

Norris, K. R. (1965). *Ann. Rev. Ent.* **10**, 47–68. The bionomics of blow-flies.

Norris, K. R. (1966). *Aust. J. Zool.* **14**, 835–853. Daily pattern of flight activity of blow-flies (*Calliphoridae: Diptera*) in the Canberra district as indicated by trap catches.

Nuorteva, P. (1959). *Ann. ent. Fen.* **25**, 1–137. Studies on the significance of flies in the transmission of poliomyelitis, I-IV.

Ovazza, M. Coz, J., and Ovazza, L. (1965). *Bull. Soc. Path. Exot.* **58**, 938–950. Etude des populations de *Simulium damnosum* Theobald, 1903 (*Diptera: Simuliidae*) en zones de gites non permanents: I. Observations sur les variations de quelques uns des caracteres utilises dans l'estimation de l'age physiologique.

Ovazza, M., Ovazza, L., and Balay, G. (1965). *Bull. Soc. Path. Exot.* **58**, 1118–1154. Etude des populations de *Simulium damnosum* Theobald, 1903 (*Diptera: Simuliidae*) en zones de gites non permanents: II. Variations saisonnieres se produisant dans les populations adultes et preimaginales.

Page, W. A. (1959a). *Bull. ent. Res.* **50**, 595–615. The ecology of *Glossina longipalpis* Wied. in Southern Nigeria.

Page, W. A. (1959b). *Bull. ent. Res.* **50**, 617–631. The ecology of *G. palpalis* (R–D) in Southern Nigeria.

Page, W. A. (1959c). *Bull. ent. Res.* **50**, 633–646. Some observations on the *fusca* group of tse-tse flies (*Glossina*) in the south of Nigeria.

Pal, Rajindar, Nair, C. P., Ramalingam, S., Patel, P. V., and Baboo Ram (1960). *Ind. J. Mal.* **14**, 595–604. The bionomics of vectors of human filariasis in Ernakulum (Kerala), India.

Parker, D. D. (1958). *J. econ. Ent.* **51**, 32–36. Seasonal occurrence of fleas on antelope ground squirrels in the Great Salt Lake desert.

Parr, M. J. (1965). *Field Studies.* **2**, 237–282. A population study of a colony of imaginal *Ischnura elegans* (van der Linden) (*Odonata: Coenagriidae*) at Dale, Pembrokeshire.

Pausch, R. D., and Provost, M. W. (1965a). *Mosquito News.* 1–8. The dispersal of *Aedes taeniorhynchus*: IV. Controlled field production.

Pausch, R. D., and Provost, M. W. (1965b). *Mosquito News.* **25**, 9–14. The dispersal of *Aedes taeniorhynchus:* V. A controlled synchronous emergence.

Peschken, D., and Thorsteinson, A. J. (1965). *Ent. expt.* applic. **8**(4), 282. Visual orientation of black-flies (*Simuliidae: Diptera*) to colour, shape and movement of targets.

Petrie, G. F., and Todd, R. E. (1923). Reports and notes of the Public Health Laboratory, Cairo, 5. Plague Report. Dept. of Public Health, Ministry of the Interior. Egypt. Govt. Press. Cairo.

Petrischeva, P. A. (1946). *J. gen. Biol.* **7**, 65–84 (in Russian). Sand-flies (*Phlebotomus*) in various landscape zones of U.S.S.R.: I. Sand-flies in hot deserts in Central Asia.

Petrischeva, P. A. (1965). Sand-flies (*Phlebotominae, Diptera, Nematocera, Psychodidae*) in Vectors of diseases of natural foci. (Ed. P. A. Petrischeva) 57–85. Israel Program for Scientific Translations. Oldbourne Press, London.

Pickens, L. G., Morgan, N. O., Hartsock, J. G., and Smith, J. W. (1967). *J. Econ. Ent.* **60**, 1250–1255. Dispersal patterns and populations of the house-fly affected by sanitation and weather in rural Maryland.

Pilson, R. D., and Leggate, B. M. (1962a). *Bull. ent. Res.* **53**, 541–549. A diurnal and seasonal study of the feeding activity of *Glossina pallidipes* Aust.

Pilson, R. D., and Leggate, B. M. (1962b). *Bull. ent. Res.* **53**, 551–562. A diurnal and seasonal study of the resting behaviour of *Glossina pallidipes* Aust.

Pilson, R. D., and Pilson, B. M. (1967). *Bull. ent. Res.* **57**, 227–257. Behaviour studies of *Glossina morsitans* Westw. in the field.

Plague Commission (1906) *J. Hyg.* **6**, No. 4. "Plague number." 421–536. Reports on plague investigations in India.

Power, R. J. B. (1964). *Proc. R. ent. Soc. Lond.* A**39**, 5–14. The activity pattern of *Glossina longipennis Corti* (*Diptera: Muscidae*).

Provost, M. W. (1952). *Mosquito News.* **12**, 174. The dispersal of *Aedes taeniorphynchus:* I. Preliminary studies.

Provost, M. W. (1957). *Mosquito News.* **17**, 233. The dispersal of *Aedes taeniorhynchus:* II. The second experiment.

Quate, L. W. (1964). *J. med. Ent.* **1**, 213–268. Phlebotomus sand-flies of the Paloich area in the Sudan.

Quarterman, K. D., Mathis, W., and Kilpatrick, J. W. (1954). *J. econ. Ent.* **47**, 405–12. Urban fly dispersal in the area of Savannah, Georgia.

Quarterman, K. D., Kilpatrick, J. W., and Mathis, W. (1954). *J. econ. Ent.* **47**, 413–419. Fly dispersal in a rural area near Savannah, Georgia.

Rachou, R. G., Ferriera, M. O., Martins, C. M., and Neto, J. A. F. (1958). *Rev. Bras. Malar. Doc. Trop.* **10**, 51. Monthly variations in the domiciliary density of *Culex pipiens fatigans* in Florianopolis (Santa Catarina).

Rachou, R. G., Lyons, G., Moura-Lima, M., and Kerr, J. A. (1965). *Am. J. trop. Med. Hyg.* **14**, 1–62. Synoptic epidemiological studies in El Salvador.

Rahman, M. A., and Ahmad, N. (1963). *Pakistan J. Hlth.* **13**, 118–151. Flea index studies in Lahore.

Raybould, J. N. (1964). *J. econ. Ent.* **57**, 445–447. An improved technique for sampling the indoor density of African house-fly populations.

Raybould, J. N. (1966a). *J. Econ. Ent.* **59**, 639–644. Techniques for sampling the density of African house-fly populations: I. A field comparison of the use of the Scudder Grill and the sticky fly-trap method for sampling the indoor density of African house-flies.

Raybould, J. N. (1966b). *J. econ. Ent.* **59**, 644–648. Techniques for sampling the density of African house-fly populations: II. A field comparison of the Scudder Grill and the sticky fly-trap method for sampling the outdoor density of African house-flies.

Reeves, W. C., Tempelis, C. H., Bellamy, R. E., and Lofy, M. F. (1963). *Am. J. trop. Med. Hyg.* **12**, 929–935. Observations on the feeding habits of *Culex tarsalis* in Kern county, California, using precipitating antisera produced in birds.

Reid, J. A. (1961). *Bull. ent. Res.* **52**, 43–62. The attraction of Mosquitoes by human or animal baits in relation to transmission of disease.

Rennison, B. D., Lumsden, W. H. R., and Webb, C. J. (1958). *Nature.* **181**, 1354–5. Use of reflecting paints for locating tse-tse fly at night.

Reuben, R. (1963). *Proc. R. ent. Soc. Lond.* **38**, 181–193. A comparison of trap catches of *Culicoides impunctatus* Goetghebuer (*Diptera: Ceratopogonidae*) with meteorological data.

Reuben, R. (1965). *Ind. J. Mal.* **17**, 223–231. Natural mortality in mosquitoes of the *Culex vishnui* group in South India.

Robertson, D. H. H. (1962). *I.S.C.T.R. 9th Meeting, Conakry.* 337–341. Some recent research developments at the East African trypanosomiasis research organization.

Robinson, C. G. (1965). *Bull. ent. Res.* **56**, 351–355. A note on nocturnal resting sites of *Glossina morsitans* Westw. in the republic of Zambia.

Rosen, P., and Gratz, N. G. (1959). *Bull. W.H.O.* **20**, 841–847. Tests with organo-phosphorus dry sugar baits against house-flies in Israel.

Rothschild, M. (1960). *Ent. Month. Mag.* **96**, 106–109. Observations and speculations concerning the flea vector of myxomatosis in Britain.

Rothschild, M., and Ford, B. (1965a). *Proc. XIIth Int. Congr. Ent.* (London): 801–802. Reproductive hormones of the host controlling the sexual cycle of the rabbit flea *Spilopsyllus cuniculi* (Dale).

Rothschild, M., and Ford, B. (1965b). *Proc. R. ent. Soc. Lond.* **40**, 109–117. Observations on gravid rabbit fleas *Spilopsyllus cuniculi* (Dale) parasitizing the hare (*Lepus europaeus* Pallas), together with further speculations concerning the course of myxomatosis at Ashton, Northants.

Russell, P. F. (1931). *Philippine J. Sci.* **46**, 639–649. Day-time resting places of Anopheles mosquitoes in the Philippines. First report.

Russell, P. F., and Santiago, D. (1934). *Proc. ent. Soc. Wash.* **36**, 1–21. An earth-lined trap for Anopheline mosquitoes.

Samarawickrema, W. A. (1962). *Ann. trop. Med. Parasit.* **56**, 110–126. Changes in the ovariole of Mansonia (Mansonioides) mosquitoes in relation to age determination.

Samarawickrema, W. A. (1967). *Bull. W.H.O.* **37**, 117–137. A study of the age composition of natural populations of *Culex pipiens fatigans* Wiedmann in relation to the transmission of filariasis due to Wuchereria bancrofti (Cobbald) in Ceylon.

Saunders, D. S. (1960). *Nature.* **186**, 651. Determination of physiological age of female *Glossina morsitans*.

Saunders, D. S. (1962). *Bull. ent. Res.* **53**, 579–595. Age determination for female tse-tse flies and the age composition of samples of *Glossina pallidipes* Aust. *G. palpalis fuscipes* Newst., and *G. brevipalpis* Newst.

Saunders, D. S. (1964). *Bull. ent. Res.* **55**, 483–497. The effect of site and sampling method on the size and composition of catches of tse-tse flies, (Glossina) and Tabanidae (Diptera).

Scherer, W. F., and Buescher, E. L., (1959). *Am. J. trop. Med. Hyg.* **8**, 644–650. Ecological studies of Japanese encephalitis in Japan: I. Introduction.

Scherer, W. F., Buescher, E. L., Flemings, M. B., Noguchi, A., and Scanlon, J. (1959). *Am. J. trop. Med. Hyg.* **8**, 665–677. Ecological studies of Japanese encephalitis virus in Japan: III. Mosquito factor, zootropism and vertical flight of

Culex tritaeniorhynchus with observations on variations in collections from animal baited traps in different habitats.

Schoof, H. F. (1951). *Summary investig. No. 27. C.D.C. Technology Branch. U.S. Dept. Hlth, Education Welfare*. Experimental design for study of the relationship between grill counts, bait-trap catches, and population levels of *M. domestica, P. pallescens* and *C. macellaria*

Schoof, H. F. (1955). *Publ. Hlth. Monograph No. 32*. Publ. Hlth. Reports. Wash. Survey and appraisal methods for community fly-control programmes.

Schoof, H. F. (1967). *Bull. W.H.O.* **36**, 600–601. Mating, Resting Habits and Dispersal of *Aedes aegypti*.

Scott, D. (1960). *C.C.T.A. 8th Meeting, Jos, Nigeria*, **45**. A recent study of outbreaks of human trypanosomiasis in Northern Ghana (1957–59).

Scudder, H. L. (1947). *U.S. Publ. Hlth. Reports*. **62**, 681–686. A new technique for sampling the density of house-fly populations.

Seal, S. C. (1960). *Bull. W.H.O.* **23**, 293–300. Epidemiological studies of plague in India: 2. The changing pattern of rodents and fleas in Calcutta and other cities.

Senior White, R. A. (1951). *Ind. J. Mal.* **5**, 465–512. Studies on the bionomics of *Anopheles aquasalis* Curry 1932. Part II.

Senior White, R. A. (1952). *Ind. J. Mal.* **6**, 29–72. Studies on the bionomics of *Anopheles aquasalis* Curry. Part III.

Service, M. W. (1963). *Bull. ent. Res.* **54**, 601–632. The ecology of the mosquitoes of the northern Guinea savannah of Nigeria.

Service, M. W. (1964). *J. ent. Soc. S. Africa*. **1**, 29–36. The attraction of mosquitoes by animal baits in the Northern Guinea Savannah of Nigeria.

Service, M. W., and Boorman, J. P. T. (1965). *Cahiers de O.R.S.T.O.M. Entomologie medicale*. Nos. **3**, **4**, 27–33. An appraisal of adult mosquito trapping techniques used in Nigeria, West Africa.

Shortt, H. E. (1932). Indian Medical Research Memoirs. No. 25, 1–200. Reports of the Kala-Azar Commission, India. Report No. II (1926–30).

Smith, A. (1965a). *Bull. ent. Res.* **56**, 161–167. A verandah trap hut for studying the house-frequenting habits of mosquitoes and for assessing insecticides.

Smith, A. (1965b). *Bull. ent. Res.* **56**, 275–282. A verandah trap hut for studying the house-frequenting habits of mosquitoes and for assessing insecticides: II. The effect of dichlorvos (DDVP) on egress and mortality of *Anopheles gambiae* Giles and *Mansonia uniformis* Theo. entering naturally.

Smith, I. M., and Rennison, B. D. (1961). *Bull. ent. Res.* **52**, 165–189. Studies of the sampling of *Glossina pallidipes* Aust. I. The number caught daily on cattle, in Morris traps and on a fly round; II. The daily pattern of flies caught on cattle, in Morris traps and on a fly round.

Smith, W. W., and Love, G. J. (1956). *Mosquito News*. **16**, 279–281. Effects of drought on the composition of rural mosquito populations as reflected by light-trap catches.

Soerono, M., Badawi, A. S., Muir, D. A., Soedono, A., and Siran, N. (1965). *Bull. W.H.O.* **33**, 453–459. Observations on doubly resistant *Anopheles aconitus* Donitz in Java, Indonesia, and on its amenability to treatment with malathion.

Soper, F. J., Bruce Wilson, D., Servulo Lima, and Waldemar sa Antunes (1943). In *The organization of permanent nation wide anti-Aedes measures in Brazil*. 137 pp. The Rockefeller Foundation, New York.

Southon, H. A. W., (1958). *CCTA. Int. Sci. Comm. Tryp. Res. 7th Meeting. Brussels* Night observations on *G. swynnertoni* Austen.

Southwood, T. R. E. (1966). Ecological methods with particular reference to the study of insect populations. London. 391 pp. Methuen and Co., London.

Standfast, H. A. (1965). *Mosquito News*. **25**, 48–53. A miniature light trap which automatically segregates the catch into hourly samples.

Stewart, M. A., and Evans, F. C. (1941). *Proc. Soc. expt. Biol. Med.* N.Y. **47**, 140. A comparative study of rodent and burrow-flea populations.

Symes, C. B. (1960). *J. trop. Med. Hyg.* **63**, Nos. 1–3. 1–14, 31–44, 59–67. Observations on the epidemiology of filariasis in Fiji.

Taylor, L. R. (1951). *Ann. appl. Biol.* **38**, 582–591. An improved suction trap for insects.

Taylor, R. M., Work, T. H., Hurlbut, H. S., and Farag Rizk (1956). *Am. J. trop. Med. Hyg.* **5**, 579–620. A study of the ecology of West Nile virus in Egypt.

Teesdale, C. (1955). *Bull. ent. Res.* **46**, 711–742. Studies on the bionomics of *Aedes aegypti* (L) in its natural habitats in a coastal region of Kenya.

Tempelis, C. H., Francy, D. B., Hayes, R. O., and Lofy, M. F. (1967). *Am. J. trop. Med. Hyg.* **16**, 111–119. Variations in feeding patterns of seven culicine mosquitoes on vertebrate hosts in Weld and Larimer counties, Colorado.

Tempelis, C. H., Reeves, W. C., Bellamy, R. E., and Lofy, M. F. (1965). *Am. J. trop. Med. Hyg.* **14**, 170–177. A three-year study of the feeding habits of *Culex tarsalis* in Kern County, California.

Thatcher, V. E., and Hertig, M. (1960). *Ann. ent. Soc. Am.* **59**, 46–52. Field studies on the feeding habits and diurnal shelters of some Phlebotomus sand-flies (*Diptera: Psychodidae*) in Panama.

Thorsteinson, A. J., Bracken, G. K., and Hanec, W. (1965). *Ent. expt. appl.* **8**, 189–192. The orientation behaviour of horse-flies and deer-flies (*Tabanidae, Diptera*) III. The use of traps in the study of orientation of Tabanids in the field.

Tinker, M. E. (1967). *J. econ. Ent.* **60**, 634–637. Measurement of *Aedes aegypti* populations.

Townes, H. K. (1962). *Proc. ent. Soc. Wash.* **64**, 253–262. Design for a Malaise trap.

Turner, E. R., and Hoogstraal, H. (1965). *J. med. Ent.* **2**, 137–139. Leishmaniasis in the Sudan Republic, 23. Sand-flies (*Phlebotomus*) attracted to rodent baited traps (*Diptera: Psychodidae*)

van den Berghe, L., and Lambrecht, F. L. (1954). *Bull. ent. Res.* **45**, 501–505. Notes on the discovery of *Glossina brevipalpis* Newst. in the Mosso Region (Urundi).

van den Berghe, L., and Lamprecht, F. L. (1962). *Academic royale des sciences d'outres mer. Memoirs. Tome XIII. fasc.* **4**, 116. Etude biologique et ecologique de *Glossina morsitans* Westw. dans la region de Bugesera (Rwanda).

van den Berghe, L., and Lambrecht, F. L. (1963). *Am. J. trop. Med. Hyg.* **12**, 129–164. The epidemiology and control of human trypanosomiasis in *Glossina morsitans* fly belts.

Wattal, B. L., and Kalra, M. L. (1960). *Ind. J. Mal.* **14**, 605–616. Studies on culicine mosquitoes: I. Preferential indoor resting habits of *Culex fatigans* Wiedmann 1828 near Ghaziabad, Uttar Pradesh.

Weitz, B. (1956). *Bull. W.H.O.* **15**, 473. Identification of blood meals of blood-sucking arthropods.

Weitz, B. (1963). *Bull. W.H.O.* **28**, 711–729. The feeding habits of *Glossina*.

Weitz, B., and Glasgow, J. P. (1956). *Trans. R. Soc. trop. Med. Hyg.* **50**, 593. The natural hosts of some species of Glossina in East Africa.

Welch, S. F., and Schoof, H. F. (1953). *Am. J. trop. Med. Hyg.* **2**, 1131–1136. The reliability of "visual survey" in evaluating fly densities for community control programmes.

Wenk, P. (1965). *Z. Morph. Okol. Tiere.* **55**, 656–670. Uber die Biologie Blutsaugender Simuliiden (Diptera) I. Besamungsrate der beim Blutenbesuch und Anflung auf den Blutwirt.

Wenk, P., and Schlorer, G. (1963). *Zeit, trop. Med. Parasit.* **14**, 177–191. Wirtsorientierung and Kopulation bei blutsaugenden Simuliiden (Diptera).

Wharton, R. H. (1951). *Ann. trop. Med. Parasit.* **45**, 141–154. The habits of adult mosquitoes in Malaya: I. Observations on anophelines in window-trap huts and in cattle sheds; *Ann. trop. Med. Parasit.* **45**, 155–160. II. Observations on culicines in window-trap huts and at cattle sheds.

Wharton, R. H. (1959). *Nature. Lond.* **184**, 830. Age determination in Mansonioides mosquitoes.

Wharton, R. H. (1962). *Bull.* No. **11.** *Inst. Med. Res. Fed. Malaya.* 114. The biology of Mansonia mosquitoes in relation to the transmission of filariae in Malaya.

Wharton, R. H., Eyles, E., and Warren, M. W. (1963). *Ann. trop. Med. Parasit.* **57**, 32–46. The development of methods for trapping the vectors of monkey malaria.

Williams, C. B. (1962). *Trans. R. ent. Soc. Lond.* **114**, 28–47. Studies on black-flies (*Diptera: Simuliidae*) taken in a light trap in Scotland. Part 3. The relation of night activity and abundance to weather conditions.

Williams, C. B. (1964). Patterns in the Balance of Nature and related problems in quantitative biology. 324 pp. In *Theoretical and experimental Biology Monographs.* No. 60. Academic Press, London and New York.

Williams, C. B. (1965). *Proc. R. ent. Soc. Lond.* **40**, 92–95. Black-flies (*Diptera: Simuliidae*) in a suction trap in the central highlands of Scotland.

Williams, P. (1965). *Ann. trop. Med. Parasit.* **59**, 393–404. Observations on the Phlebotomus sand-flies of British Honduras.

Wirth, W. W., and Bottimer, L. J. (1956). *Mosquito News.* **16**, 256–266. A population study of the *Culicoides* midges of the Edwards Plateau region of Texas.

Wolfe, L. S., and Peterson, D. G. (1960). *Can. J. Zool.* **38**, 489–497. Diurnal behaviour and biting habits of black-flies (Diptera: Simuliidae) in the forests of Quebec.

W.H.O. (World Health Organization). (1963). *Practical Entomology in Malaria Eradication.* (R. C. Muirhead-Thomson, ed.). Mimeographed edition. 2 vols. 354 pp. World Health Organization, Geneva.

W.H.O. (1965) Expert Committee on Onchocerciasis. 2nd report. *W.H.O. Technical Report Series.* No. **335**, 96. Geneva.

W.H.O. World Health Organization) (1967b). Mosquito Ecology. Report of a W.H.O. Scientific group. *W.H.O. Technical series* No. **368**. 22 pp. Geneva.

W.H.O. (1967a). *Bull. W.H.O.* **36**, 519–702. Seminar on the Ecology, Biology, Control and Eradication of *Aedes aegypti.*

AUTHOR INDEX

Numbers in italics indicate those pages in the Bibliography where references are listed.

SUBJECT INDEX